HANDBOOK OF LABORATORY ANIMAL ANESTHESIA and PAIN MANAGEMENT
RODENTS

The **Laboratory Animal Pocket Reference** Series

Series Editor
Mark A. Suckow, D.V.M.
Freimann Life Science Center
University of Notre Dame
South Bend, Indiana

Published Titles

Handbook of Laboratory Animal Anesthesia and
Pain Management: Rodents

The Laboratory Bird

Critical Care Management for Laboratory Mice and Rats

The Laboratory Canine

The Laboratory Cat

The Laboratory Ferret

The Laboratory Guinea Pig, Second Edition

The Laboratory Hamster and Gerbil

The Laboratory Mouse, Second Edition

The Laboratory Nonhuman Primate

The Laboratory Rabbit, Second Edition

The Laboratory Rat, Second Edition

The Laboratory Small Ruminant

The Laboratory Swine, Second Edition

The Laboratory *Xenopus sp.*

The Laboratory Zebrafish

A Volume in The Laboratory Animal Pocket Reference Series

HANDBOOK OF LABORATORY ANIMAL ANESTHESIA and PAIN MANAGEMENT
RODENTS

Edited by
Cholawat Pacharinsak
Director of Anesthesia, Pain Management, and Surgery
Department of Comparative Medicine, Stanford University
School of Medicine, California, USA

Jennifer C. Smith

CRC Press
Taylor & Francis Group
Boca Raton   London   New York

CRC Press is an imprint of the
Taylor & Francis Group, an **informa** business

CRC Press
Taylor & Francis Group
6000 Broken Sound Parkway NW, Suite 300
Boca Raton, FL 33487-2742

Printed on acid-free paper
Version Date: 20160725

International Standard Book Number-13: 978-1-4665-8567-6 (Paperback)

Library of Congress Cataloging-in-Publication Data

Names: Pacharinsak, Cholawat, author. | Smith, Jennifer C. (Jennifer Capra),
1967- author.
Title: Handbook of laboratory animal anesthesia and pain management : rodents
/ Cholawat Pacharinsak and Jennifer C. Smith.
Other titles: Laboratory animal pocket reference series.
Description: Boca Raton : Taylor & Francis, 2017. | Series: Laboratory animal
pocket reference series | Includes bibliographical references and index.
Identifiers: LCCN 2016032842| ISBN 9781466585676 (pbk. : alk. paper) | ISBN
9781466585683 (alk. paper)
Subjects: | MESH: Anesthesia--veterinary | Animals, Laboratory | Rodentia |
Pain--prevention & control | Anesthetics | Handbooks
Classification: LCC SF914 | NLM QY 39 | DDC 636.089/796--dc23
LC record available at https://lccn.loc.gov/2016032842

Visit the Taylor & Francis Web site at
http://www.taylorandfrancis.com

and the CRC Press Web site at
http://www.crcpress.com

I would like to dedicate this work to my parents, Anck and Pattraporn Pacharinsak.

Cholawat Pacharinsak

To my mom, Lucille Capra, who continually shares her passion for the written word.

Jennifer C. Smith

contents

preface

Rodents are the most commonly used species in biomedical research. In many instances, individuals performing such research are charged with broad responsibilities, including the performance of technical procedures that directly or indirectly require anesthesia or pain management. Therefore, the animals' comfort is a critical component of conducting ethical, compassionate, scientifically valid biomedical research. Taken together or individually, anesthesia and pain management are complex concepts involving several components (anesthetic machine/equipment, anesthetic monitoring, pharmacology of anesthetics or analgesics, pain management, etc.) toward producing sound, safe, and effective anesthesia and pain management. This complexity can challenge researchers, scientists, veterinarians, or technicians. This book emphasizes rodent anesthesia and pain management in the research setting, including understanding the anesthetic machine and related equipment, anesthetic management and monitoring, anesthesia and analgesia pharmacology, euthanasia, and record keeping. The format is different than other textbooks, as bullet points are used to facilitate navigation through each chapter. In this regard, this handbook was written to provide a quick reference source for investigators, technicians, and animal caretakers charged with the care and/or use of rodents in a research setting. It should be particularly valuable to those at small institutions or facilities lacking a large, well-organized animal resource unit and to those individuals who need to conduct research programs using rodents starting from scratch.

This handbook is organized into nine chapters: (1) terms and definitions; (2) understanding anesthetic-related equipment; (3) anesthetic pharmacology and research-related anesthetic considerations;

(4) management of anesthesia; (5) anesthetic monitoring; (6) special techniques and species; (7) pain management; (8) euthanasia; and (9) regulatory management of rodent anesthesia. Basic information and common procedures are presented in detail. Other information regarding alternative techniques and details of procedures and methods beyond the scope of this handbook is referenced extensively, so users are directed toward additional information without having to wade through a burdensome volume of detail. In this sense, this handbook should be viewed as a basic reference source and not as an exhaustive review of rodent anesthesia and pain management.

A final point for consideration is that all individuals performing procedures requiring anesthesia and analgesia must be properly trained. The humane care and use of rodents is improved by initial and continued education of personnel and will facilitate the overall process of programs using rodents in research, teaching, and testing.

Cholawat Pacharinsak
Jennifer C. Smith

acknowledgments

The editors wish to acknowledge Ms. Janis Atuk-Jones, who provided significant assistance, and Dr. Patrick Sharp, who provided advice on the book-editing process. We would like to thank all personnel and/or companies who provided us with graphics or images.

editors

Cholawat Pacharinsak is an assistant professor and director of Veterinary Anesthesia, Pain Management, and Surgery at Stanford University's Department of Comparative Medicine. He is also a Diplomate of the American College of Veterinary Anesthesia and Analgesia (DACVAA). Dr. Pacharinsak received his Doctor of Veterinary Medicine (DVM) degree from Chulalongkorn University, Thailand, trained in an anesthesiology and pain management residency program, and received his Master's degree from Washington State University. He completed his PhD in pain neuroscience from the Comparative Molecular Biosciences program at the University of Minnesota. Prior to arriving at Stanford, Dr. Pacharinsak was a faculty member in anesthesiology and pain management at Michigan State University and Purdue University, and he served as a clinical specialist at UCLA.

Jennifer C. Smith is the director of Bioresources and the attending veterinarian at the Henry Ford Health System in Detroit, Michigan. She is a Diplomate of the American College of Laboratory Animal Medicine (ACLAM). She received her DVM from Michigan State University College of Veterinary Medicine, where she also received her residency training in veterinary anesthesia; prior to her directorship at Henry Ford, Dr. Smith provided veterinary oversight, management, and research support to large vendor, pharmaceutical, and biotechnology companies. She is an active volunteer with the American Association For Laboratory Animal Science (AALAS) and also serves as an ad hoc consultant to the Assessment and Accreditation of Laboratory Animal Care—International AAALAC.

contributors

Patricia Foley is director of the Division of Comparative Medicine and professor at the Department of Microbiology and Immunology at Georgetown University. She received her DVM from the University of California at Davis in 1986. Dr. Foley is a Diplomate of the American College of Laboratory Animal Medicine, and is a Certified Professional IACUC [Institutional Animal Care and Use Committee] Administrator (CPIA). Following 5 years of private practice, Dr. Foley has authored or coauthored over 30 peer-reviewed publications and has conducted funded research on analgesics and physiological monitoring devices, with a particular interest in sustained release analgesia.

Janlee Jensen is a veterinary specialist based in Los Angeles, CA.

Robert D. Keegan is an associate professor and anesthesia section head at the Washington State University College of Veterinary Medicine. Dr. Keegan completed his residency in veterinary anesthesia at Cornell University and is a Diplomate of the American College of Veterinary Anesthesia and Analgesia.

C. Tyler Long is a clinical veterinarian at North Carolina State University College of Veterinary Medicine. He is a Diplomate of the American College of Laboratory Animal Medicine and American Board of Toxicology. He received his DVM from Ross University and spent 4 years in private practice and teaching in the veterinary technician program at Stautzenberger College prior to returning to academia to complete a 3-year residency at Stanford University Medical Center in Palo Alto, California. After his residency, he worked as a clinical veterinarian and surgeon for a private contract research organization specializing in medical device testing before returning to his home state of North Carolina in 2013.

Daniel Pang is an associate professor in the Department of Clinical Sciences of the Faculty of Veterinary Medicine at the University of Montreal. He is a Diplomate of the American and European Colleges of Veterinary Anaesthesia and Analgesia. After graduation from the University of Bristol veterinary school in the United Kingdom, he worked briefly in private practice before undertaking an internship (small animal) at the University of Glasgow. This was followed by a combined MSc-residency training program at the University of Montreal and a PhD on the mechanisms of action of volatile anesthetics at Imperial College, London. After returning to Canada, Dr. Pang was a faculty member at the University of Calgary from 2010 to 2016 before joining the University of Montreal.

Travis Seymour is a clinical veterinarian at The Children's Hospital of Philadelphia. He received his Veterinariae Medicinae Doctoris (VMD) degree from the University of Pennsylvania and his Master of Laboratory Animal Science degree from Drexel University. He completed the Laboratory Animal Medicine residency program at Stanford University and is a Diplomate of the American College of Laboratory Animal Medicine.

Patrick Sharp is the chief executive officer of Western Australia's Animal Resources Authority. He received his DVM degree from Purdue University and completed his postdoctoral fellowship at the Washington University School of Medicine in St. Louis. Dr. Sharp serves as a private consultant for various organizations providing various services including facility planning/design and AAALAC accreditation preparation. Dr. Sharp has worked in government, industry, and academia in the United States, Asia, Europe, and Australia and has served as an AAALAC ad hoc consultant specialist. Dr. Sharp supports comparative medicine training, including both American College of Laboratory Animal Medicine (ACLAM)-approved training programs and international training opportunities for laboratory animal graduate veterinarians and veterinary students. He strives to improve the quality of animal care through personnel education at all animal care levels and through the design and construction of better vivaria and research support facilities.

Andre Shih graduated from the University of Sao Paulo Brazil in 1999 with a degree in veterinary medicine. He moved to the United States in that year and finished his DVM training at the University of Wisconsin—School Veterinary Medicine in Madison. Upon graduation, he worked as an emergency and critical care veterinarian in Fox Valley Animal Referral Center in Wisconsin for a couple of years. He completed his anesthesia residency at the University of Florida where he is currently an associate professor of anesthesia. He has a certificate in acupuncture and has just finished his Emergency Critical Care Fellowship residency. He is also board certified by both the American College of Veterinary Emergency Critical Care and the American College of Veterinary Anesthesia and Analgesia. His areas of research include preload monitoring, hypovolemic shock, and cardiopulmonary resuscitation.

terms and definitions

Travis Seymour

- *Acute pain*: Pain associated with tissue damage that resolves once the tissue heals
- *Allodynia*: Pain resulting from nonnoxious stimuli
- *Anesthesia*: Loss of sensation to the whole or any part of the body
- *Chronic pain*: Persistent pain that lasts significantly longer than the anticipated resolution time
- *General anesthesia*: Unconsciousness due to drug-induced, reversible depression of the central nervous system
- *Hyperalgesia*: Heightened pain intensity from noxious stimuli that normally cause pain
- *Neuroleptanalgesia*: A combination of opioids and sedatives/ tranquilizers that work synergistically; provides effective analgesia and sedation while decreasing the dosing require- ments of both types of drugs
- *Neuropathic pain*: Pain resulting from injury/damage of the peripheral or central nervous system
- *Nociception*: Transmission of nociceptive information to higher CNS regions
- *Nociceptors*: Receptors associated with mainly afferent neu- rons that detect noxious stimuli

- *Noxious stimuli*: Stimuli causing tissue injury or damage (categorized as mechanical, thermal, chemical, or electrical stimuli)
- *Pain*: An unpleasant sensation with an emotional component, usually associated with actual or potential tissue damage
- *Sedation*: Depression of consciousness
- *Surgical anesthesia*: The plane of anesthesia characterized by unconsciousness, muscle relaxation, and analgesia appropriate for a surgical procedure
- *Tranquilization*: A behavioral change whereby the patient is relaxed yet cognizant of its surroundings

bibliography

Pacharinsak, C. and P. Sharp. 2013. Anesthesia in nonhuman primates. In *Pocket Handbook of Nonhuman Primate Clinical Medicine*, ed. A. Courtney, 1–32. Boca Raton, FL: CRC Press.

Tranquilli, W. J. and K. A. Grimm. 2015. Introduction: Use, definitions, history, concepts, classification, and considerations for anesthesia and analgesia. In *Veterinary Anesthesia and Analgesia, the Fifth Edition of Lumb and Jones*, eds K. A. Grimm, L. A. Lamont, W. J. Tranquilli, S. A. Greens, and S. A. Robertson. 5th edition, 23–85. Ames, IA: Wiley.

understanding anesthetic-related equipment

Daniel Pang

contents

Modern anesthetic equipment is reliable when regularly maintained and used appropriately. The provision of general anesthesia to laboratory species is often associated with additional constraints specific to the research setting, including animal positioning and access and drug and equipment restrictions. As a result, the associated anesthetic risk may be increased, emphasizing the role of the anesthetist in providing a successful outcome while safeguarding welfare.

2.1 carrier gas supply

2.1.1 Carrier Gas

The *carrier gas*, typically oxygen, is the gas used to carry an inhalational anesthetic agent to the animal. It may be supplied by connecting an anesthetic machine to a pipeline source or directly to a cylinder.

2.1.2 Pipeline Oxygen

Pipeline oxygen is common in larger units/clinics where it is more convenient to have large, less portable oxygen sources housed together in a secure, remote storage room, from which supplied gas is piped to a large number of suites. The form of oxygen source depends on workplace requirements, ranging from a bank of J cylinders to liquid oxygen containers (Figure 2.1a,b).

- Pipeline oxygen is transported at a pressure of approximately 50–60 psi (345–410 kPa) to the anesthetic machine (Figure 2.2a).
- The final pressure reduction occurs at the flowmeter (see below).
- Oxygen is stored at 1900 psi in both J and E cylinders (Figure 2.2b).

(a)   (b)

Fig. 2.1 (a) Control unit for two banks of oxygen cylinders. Using dual banks allows for one bank to be replaced with full cylinders while providing a continuous supply of gas from the second bank. The switch from one bank to the other may be manual (as in this example) or automatically controlled. In either case, messages indicating a depletion in gas supply and switching of banks may be sent to a central unit in an area observed by personnel. (b) Liquid oxygen containers used to supply a large veterinary referral hospital.

- J cylinders, containing approximately 6800 L of oxygen when full, are approximately 5 feet in height and may be used to supply pipeline oxygen or can be housed in a laboratory for local use (Figure 2.2c).

2.1.3 E Cylinders

E cylinders are fitted directly to anesthetic machines and contain approximately 660 L of oxygen when full. In the United States, oxygen cylinders are usually colored green, and in Canada, green with white shoulders.

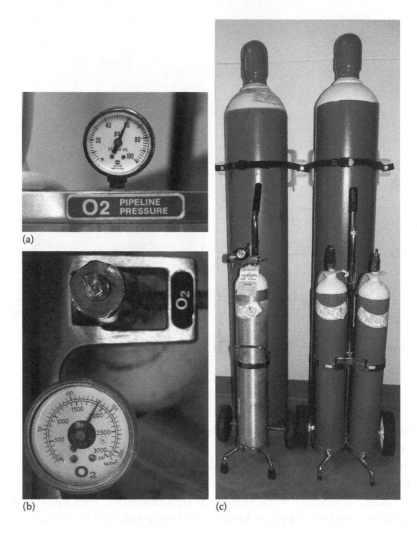

Fig. 2.2 (a) Pressure gauge showing an oxygen pipeline pressure of 59 psi at the point of connection to the anesthetic machine. (b) Pressure gauge showing an oxygen E cylinder pressure of 1900 psi, indicating a full cylinder (approximately 660 L). (c) Oxygen cylinders. Two J cylinders are in the background, with storage caps in place over valves. Notice color markings of green body with white shoulders. Both cylinders are secured to the wall with a strap and bracket. In the foreground are E cylinders stored in dollies. Notice the inconsistent color markings of the E cylinders: steel cylinders have green body with white shoulders, aluminum cylinder has silver body with white shoulders. Cylinder contents should be confirmed by reading the label (visible on E cylinders).

- To calculate the volume of oxygen remaining in an E cylinder: 0.34 × cylinder pressure (psig); for example, for a half-full cylinder, 0.34 × 950 = 323 L.
- To calculate the duration of oxygen left in an E cylinder: current cylinder volume (L, as calculated above) divided by flow rate (L/min); for example, 323 L/0.5 L/min = 646 min (a little under 11 h).

2.1.4 Pressure Regulator

Pressure reduction from cylinders to the operating pressure of an anesthetic machine or flowmeter is achieved with a pressure regulator. Pressure regulators may be nonadjustable, providing a predetermined output pressure of 60 psi; or adjustable, allowing the user to control the pressure reduction (Figure 2.3).

Fig. 2.3 Agent-specific regulator used to connect a J cylinder to an anesthetic machine. This regulator provides a fixed pressure output of 50–60 psi and has a pressure gauge showing the pressure of the cylinder contents.

2.2 gas safety

- Cylinders should be properly secured to prevent them from falling. This may be with a specifically designed cart or floor or wall bracket. When not in use, J-cylinder valves should be covered and protected with a cap (Figure 2.2c).

- Connection of E cylinders to anesthetic machines is via an agent-specific yoke. Each yoke has a hole for the passage of gas, plus two pins that correspond to holes stamped in to the E cylinder valves. This pin-hole pairing is the Pin Index Safety System (PISS), which prevents the misconnection of the wrong gas cylinder to the wrong yoke on the anesthetic machine (Figure 2.4a–d).

- Connection between a central pipeline supply and the anesthetic machine is via a flexible hose. This hose should be gas specific, using a combination of a threaded Diameter Index Safety System (DISS) connector, color coding, labeling, and often a proprietary quick connector (at the level of the wall or ceiling mount) to ensure that the correct hose is connected to the correct gas outlet (from central supply) and anesthesia machine inlet (Figure 2.4e).

- Every oxygen cylinder comes supplied with a usage label. Each label identifies the cylinder contents, is usually color-coded to match the contents, and has tear-away strips informing users of the usage state ("full," "in use," "empty"). This simple system provides rapid identification of cylinder status and is particularly useful when a cylinder is not connected to a pressure gauge to confirm filling pressure (Figure 2.4f).

2.3 the anesthetic machine

Several manufacturers provide small, relatively portable anesthetic machines specifically for laboratory animal anesthesia. These machines often include the minimum necessary equipment to provide inhalational anesthesia: (1) a single flowmeter, (2) a vaporizer, and (3) a fresh gas flow outlet for connection to a breathing system (Figure 2.5a).

(a)

(c)

(b)

(d)

Fig. 2.4 (a,b) Examples of the Pin Index Safety System for nitrous oxide. (c,d) Oxygen. Each agent has a specific pattern of blind holes in the body of the E cylinder valve (b, nitrous oxide; d, oxygen), which correspond to matching pins in the yokes on the anesthetic machine (a, nitrous oxide; c, oxygen).

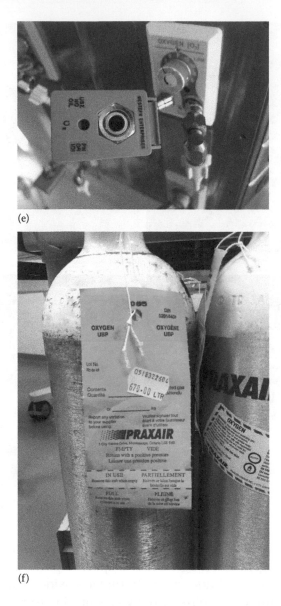

(e)

(f)

Fig. 2.4 (Continued) (e) A proprietary ceiling-mounted connection plate from which a hose can be connected to transport pipeline oxygen to an anesthetic machine. Note that the oxygen liters per minute (LPM) dial is for a separate connection designed to fit to oxygen tubing connected directly to a patient (e.g., nasal oxygen). (f) The usage label for an oxygen cylinder. It has 3 tear-away strips ("full," "in use," "empty"), which should be removed as appropriate. Notice the green color of the label, matching the color coding for oxygen.

2.4 flowmeters

Flowmeters receive gas from cylinders or a pipeline supply at approximately 50–60 psi and reduce this pressure to atmospheric pressure. They regulate gas flow through the vaporizer and to the animal. A flowmeter comprises an agent-specific, calibrated flowmeter (Thorpe) sight tube; needle valve; control knob; and float. Gas flow (mL or L/min) is calculated based on animal weight and flow requirements of the breathing system (and vaporizer).

2.4.1 Safety

- Cylindrical floats are designed to spin, confirming flow of gas (Figure 2.5b,c).
- The flow rate is read from the center of a ball float and the upper edge of a cylindrical float (Figure 2.5d).
- Control knob for oxygen is green (United States), knurled, and has a greater diameter than knobs for other gases (Figure 2.5c).
- If flowmeters for gases other than oxygen are present, they will generally be located to the left of the oxygen flowmeter to minimize gas contamination should a leak occur within the flowmeter bank.
- The control knob for nitrous oxide (N_2O), if present, may be mechanically linked to that of oxygen to prevent inadvertent administration of a hypoxic mixture.
- Flowmeter tubes do not have a constant orifice diameter for gas flow. Therefore, gas flows cannot be accurately read between or below tube markings.

2.5 anesthetic vaporizers

- Used to deliver a precise concentration of volatile anesthetic agent to a carrier gas, usually oxygen, which is then delivered to the animal. Carrier gas entering a vaporizer is split between a vaporizing chamber, where it collects anesthetic vapor; and a bypass chamber. The emerging gas plus anesthetic vapor matches the calibrated concentration dial setting.
- Modern vaporizers are temperature compensated and variable bypass. Temperature compensation is necessary as agent vaporization results in cooling of the liquid, affecting further

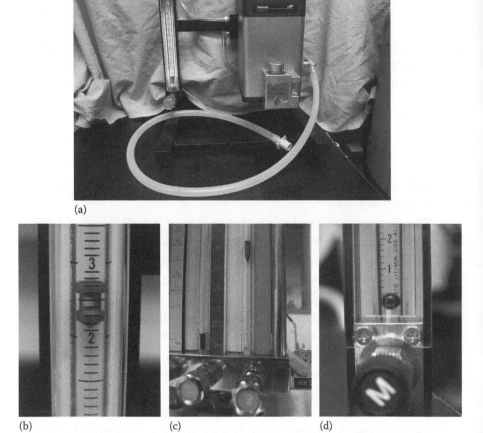

(a)

(b)   (c)   (d)

Fig. 2.5 (a) Simple tabletop anesthetic machine designed for laboratory use (Dispomed). It comprises an oxygen flowmeter, Selectatec mounted vaporizer (screw-fill), and tubing connected to fresh gas flow outlet, which allows it to function as a T-piece breathing system or to be connected to a Bain breathing system. (b) Flowmeter with a float. The white spot is used as a visual indicator that the float is spinning, confirming oxygen flow. (c) Flowmeter with float. This float does not spin. Notice the large-diameter, fluted, color-coded adjustment knob for oxygen. The nitrous oxide knob is small and smooth. (d) Flowmeter with ball float.

vaporization. Adjustment of the calibrated concentration dial controls the flow of carrier gas from the flowmeter through the vaporizer and bypass chambers (variable bypass). Within normal working temperatures and pressures, the gas emerging from the vaporizer will match the concentration dial setting.

- Vaporizers may be mounted on an anesthetic machine via either a Selectatec system or a cage mount. The Selectatec system confers the advantages of allowing tool-free rapid removal and replacement of vaporizers (Figure 2.6a).

- Vaporizers are most commonly filled using screw-fill or key-fill systems. Sevoflurane and desflurane are the exception, being supplied in sealed bottles with a dedicated vaporizer connection fitted. Screw-fill systems are the simplest to use, simply requiring the filling port to be unscrewed and the agent to be poured directly into the vaporizer. As a result, they lead to the greatest workplace pollution, and a vaporizer can easily be filled with the wrong anesthetic agent. Vaporizers can still be ordered with a screw-fill system, though they have largely been replaced by the key-fill system. This system has the added benefit of two distinct safety features designed to minimize the risk of filling a vaporizer with the wrong agent. A key-fill vaporizer is filled using an agent-specific connecting tube between the bottle of anesthetic agent and vaporizer. The distal end of the tubing is shaped to fit the correct vaporizer and the proximal end is shaped to fit the correct bottle (Figure 2.6b,c).

- Inhalational anesthetic potency is expressed by the minimum alveolar concentration (MAC). This is the alveolar anesthetic concentration that prevents movement in 50% of animals when exposed to a noxious stimulus (at sea-level atmospheric pressure). Values of MAC for a given anesthetic are fairly consistent between mammalian species. MAC is influenced by a range of factors (Table 2.1).

- To minimize the risk of an animal moving in response to noxious stimulation, a general rule of thumb is to deliver 1.3 times the anesthetic's MAC value, which prevents movement in 95% of anesthetized animals. For example, for isoflurane, this would be $1.3 \times 1.4\% = 1.8\%$.

- Potent sedatives and analgesics, such as $\alpha 2$ adrenergic agonists and opioids, can reduce MAC significantly (*MAC sparing*) (Table 2.2).

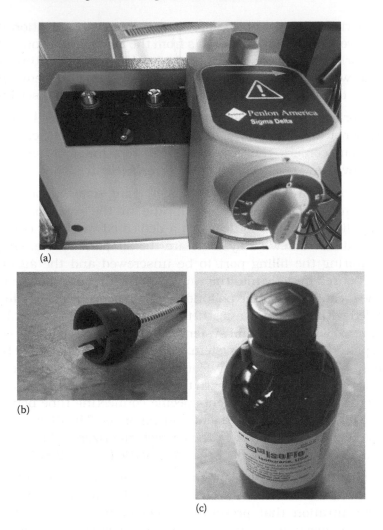

Fig. 2.6 (a) Anesthetic machine capable of having two vaporizers mounted simultaneously. The left-hand mount is unoccupied, showing the Selectatec mounting system. Vaporizer fitting and removal is accomplished by twisting the lever on top of the vaporizer to lock/unlock the vaporizer. (b) The proximal end of the connecting tube used to fill a vaporizer. The cutaway slots in the base are specific to the matching agent-specific bottle. The distal end (not shown) is shaped to fit the matching agent-specific vaporizer. (c) A bottle of isoflurane with agent-specific collar, the shape of which will only fit the matching connecting tube.

Table 2.1: COMMON FACTORS INFLUENCING MINIMUM ALVEOLAR CONCENTRATION (MAC)

Factors Increasing MAC	Factors Decreasing MAC
Catecholamines and sympathomimetics (endogenous and exogenous)	Hypoxemia
Hyperthermia	Hypothermia
Hypernatremia	Hypercapnia ($PaCO_2 > 95$ mmHg)
	Sedatives
	Injectable anesthetic agents
	Analgesics
	Pregnancy
	Hypotension

Table 2.2: MINIMUM ALVEOLAR CONCENTRATION (MAC) VALUES FOR DIFFERENT INHALATIONAL ANESTHETICS IN RATS AND MICE

Inhalational Anesthetic	Rat	Mouse
Isoflurane MAC (%)	1.38	1.41
Sevoflurane MAC (%)	2.7	2.5
Desflurane MAC (%)	5.7	6.5
$1.3 \times$ Isoflurane MAC (%)	1.8	1.8
$1.3 \times$ Sevoflurane MAC (%)	3.5	3.3
$1.3 \times$ Desflurane MAC (%)	7.4	8.5

Note: MAC multiples calculated to prevent movement in 95% of animals.

2.6 O₂ flush valve

- Depressing the valve flushes oxygen into the breathing system at a pressure of 50–60 psi (35–75 L/min) as it bypasses the flowmeter and vaporizer.
- The O_2 flush valve should not be used while an animal is connected to the breathing system due to the risk of generating a high intrathoracic pressure, barotrauma, pneumothorax, and cardiovascular collapse.

2.7 carbon dioxide (CO₂) absorber

- Essential for the normal function of a circle system as it removes CO_2 through a chemical reaction, allowing the rebreathing of exhaled gas. CO_2 reacts with water and a hydroxide base (primarily calcium hydroxide, with smaller amounts of sodium and potassium hydroxide), resulting in the formation of calcium carbonate and water, producing heat (exothermic) and a change in pH.

- The change in pH is exploited, allowing a pH-sensitive color indicator to be used to identify when the absorber is exhausted (absorptive capacity attained). The indicator color varies with each manufacturer, making it important to be aware of which color change to expect.

- Following a period of disuse, the color change will revert to its preuse color. However, it will rapidly return to the previous color at its next use.

- When a single canister is used, the absorber should be changed when the color change occupies two-thirds of the absorber volume.

- If unused for a long period, or following continuous flow of oxygen through the canister (e.g., if oxygen flowmeter is inadvertently left on overnight), the absorber granules will become desiccated. This promotes the formation of carbon monoxide, and the absorber should be changed if either of these situations occurs.

2.8 breathing systems

2.8.1 Classification

The simplest classification system for modern equipment is *non-rebreathing* or *rebreathing*. A small number of breathing systems are in common use: circle system (rebreathing), T-piece (non-rebreathing), and Bain (non-rebreathing).

2.8.1.1 Non-rebreathing systems

Non-rebreathing systems (commonly used in rodent anesthesia)

2.8.1.1.1 Bain system (coaxial Mapleson D)

- A Bain breathing system has a coaxial tube arrangement with a smaller diameter tube carrying fresh gas flow within a larger corrugated hose. The system is completed with an adjustable pressure-limiting (APL) valve and reservoir bag.

- Recommended fresh gas flow rates are in the range of 100–300 mL/kg/min. However, due to the tapered sight tube, it is unsafe to try estimating flow settings on flowmeters that are not calibrated with visible gradations at low flow rates. As a result, minimum flow rates are usually limited by the lowest marked value.

- The required flow rate can be more accurately determined by measuring CO_2 in an arterial blood sample ($PaCO_2$) or end-tidal CO_2 via capnography and decreasing the oxygen flow rate to a level that prevents rebreathing.

- With frequent use, over time the inner tube can become detached, leading to a large increase in dead space and rebreathing of CO_2.

2.8.1.1.2 T-piece (Figure 2.7a,b)

- T-piece breathing systems are examples of Mapleson E and F systems. Compared to other non-rebreathing systems, the T-piece does not include an APL valve, reducing the resistance to breathing.

- The least efficient of the non-rebreathing systems, with recommended flow rates of 300–400 mL/kg/min.

- Mapleson E and F systems differ in the absence (Mapleson E, also known as an Ayre's T-piece) or presence (Mapleson F, also known as a Jackson-Rees T-piece) of a reservoir bag.

2.8.1.2 Rebreathing systems: Circle system

Circle systems are most simply classified as rebreathing due to the presence of a carbon dioxide absorber. This allows the recirculation of exhaled gas, minus CO_2, with the addition of fresh gas (oxygen and volatile anesthetic agent).

2.8.1.2.1 Circle-system components

- One-way (unidirectional) valves
- An adjustable pressure-limiting (APL, "pop-off") valve
- Carbon dioxide absorber
- Reservoir bag
- Breathing hoses (one inspiratory, one expiratory)
- The resistance from the one-way valves and carbon dioxide absorber is associated with an increased work of breathing, limiting its use in rodents.

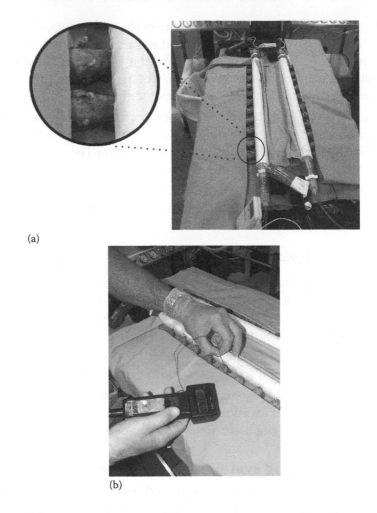

(a)

(b)

Fig. 2.7 (a) An ingenious modification of a T-piece breathing system, allowing simultaneous general anesthesia of 10 mice. The mice are connected to white plastic tubing with predrilled holes, each covered with a rubber diaphragm to provide a tight seal around the animal's nose. Anesthetic gas flows in one direction from the anesthetic machine and is scavenged at the other end. Photo provided by Professor Paul Flecknell, University of Newcastle (United Kingdom). (b) Simple monitoring of body temperature from the oropharynx. Photo provided by Professor Paul Flecknell, University of Newcastle (United Kingdom).

Table 2.3: COMPARISON OF THE ADVANTAGES AND DISADVANTAGES OF
REBREATHING AND NON-REBREATHING SYSTEMS

Rebreathing System		Non-Rebreathing System	
Advantages	Disadvantages	Advantages	Disadvantages
Economical (low O_2 requirement)	Increased resistance to breathing (unidirectional valves, CO_2 absorber)	Low resistance to breathing	More costly (increased oxygen and anesthetic agent use)
Supports warm, humid gas (from rebreathing exhaled gas)	Slower changes in anesthetic depth	Rapid changes in anesthetic depth	Cold, dry gas
Less environmental pollution with anesthetic agent			More environmental pollution with anesthetic agent

- Advantages and disadvantages of rebreathing and non-rebreathing systems are shown in Table 2.3.

2.8.1.3 Open systems

- Simple to use but increase the risk of workplace pollution and severely limit control over anesthetic depth.
- Whenever possible, this technique should be employed within a fume hood to limit occupational exposure.
- Most commonly employed as a euthanasia technique in individual animals as they cause rapid loss of consciousness. However, time to death may be prolonged and a definitive secondary method (e.g., decapitation, cervical dislocation) should be considered.
- Overdose of inhalational anesthetic is currently considered an acceptable euthanasia method (with conditions) by the American Veterinary Medical Association (AVMA), and as an acceptable method (followed by another method to ensure death) by the Canadian Council on Animal Care (CCAC).
- Examples: induction chamber, mask, open drop technique.

2.8.1.3.1 Anesthetic Induction Chamber

- Induction chambers are commonly used to induce anesthesia in rodents, and a variety of designs are available.
- Anesthetic gases are denser than room air; as a result, chamber design should include inlet ports for the inflow of

anesthetic gas at the bottom of the chamber and an outlet port for the outflow of waste gas at the top of the chamber. This design promotes rapid induction of anesthesia.

- Additional design considerations are the use of a clear material (often perspex) to allow visual observation of the animal during induction of general anesthesia, and choice of a material that is easy to clean and disinfect without reacting with common cleaning materials. Examples of a single induction chamber and multichamber induction set-up are shown in Figure 2.8a,b, respectively.

2.8.1.3.2 Anesthetic Masks

- Two types of masks are commonly available, with and without accompanying scavenging tubing.

- Simple masks without scavenging are often those used for veterinary pediatric/cat anesthesia. They should be used with a close-fitting diaphragm to create a reservoir of oxygen and anesthetic agent within the mask. A separate scavenging system is necessary to remove waste gas.

- Masks specifically for rodents comprise an inner chamber through which fresh gas flows and into which a rodent's nose is placed, surrounded by an outer chamber from which waste gas is removed.

2.8.1.3.3 Open Drop Techniques

In a laboratory animal environment, an *open* system classification applies when a volatile anesthetic agent is added directly to the bottom of a chamber containing the animal. Animals should not be in direct contact with anesthetic agent. This technique is also known as *open-drop*.

2.9 reservoir bags

- Reservoir bags are found on circle and Bain breathing systems and on the Jackson-Rees modification of a T-piece system.

- They act as a reservoir, allowing positive pressure ventilation and meeting peak-flow demands during inhalation.

- Small volume bags of 250 mL are suitable for use in rodent anesthesia. Smaller bags can also be made from party balloons.

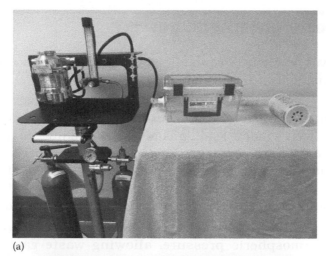

(a)

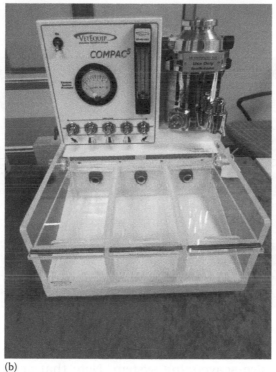

(b)

Fig. 2.8 (a) A single induction chamber. (Courtesy of Cholawat Pacharinsak.) (b) A multichamber induction. (Courtesy of Vet Equip® Inhalation Anesthesia Systems.)

2.10 waste anesthetic gas

The removal of waste anesthetic gas reduces workplace pollution. Available methods are classified as active or passive.

- *Passive scavenging systems* can vary in complexity from passing the waste gas tubing through a simple hole in a wall or an open window leading outside the building, to activated charcoal canisters (Figure 2.9a,b). Activated charcoal canisters have a maximum capacity. As they absorb volatile anesthetic agent, they increase in weight, and this weight change is monitored to ensure that the capacity is not exceeded.

- *Active scavenging systems* consist of a pump to generate a subatmospheric pressure, allowing waste gases to be

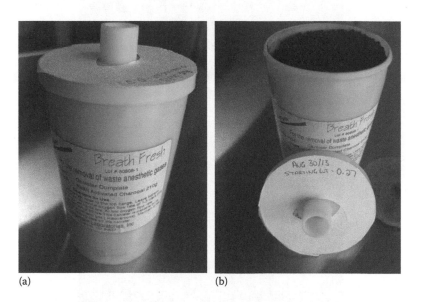

(a)   (b)

Fig. 2.9 (a) A small refillable canister containing activated charcoal, used as a passive scavenging system. Note that activated charcoal does not scavenge nitrous oxide. (b) Contents of canister showing activated charcoal granules. The starting weight of the canister with fresh granules is marked on the lid, allowing simple monitoring of a weight increase, indicating when the granules should be replaced. A thin filter (bottom right) is placed between the top of the granules and the canister lid, protecting the patient from inhalation of particles.

actively drawn away from the animal. While these are typically incorporated in to a central system connected directly to the breathing system, a more common approach in rodent anesthesia is the use of a separate system that can be easily moved. This flexibility is useful when multiple procedures performed on an animal necessitate moving the animal among several locations (e.g., adjacent counter tops) (Figure 2.10). The outlet of an active system should be vented directly outside or into a ventilation system that does not recirculate air between rooms. The use of a fume hood is a common method to achieve this. The capacity of an active scavenging system is determined by need but should be determined by expected peak flows during anesthesia.

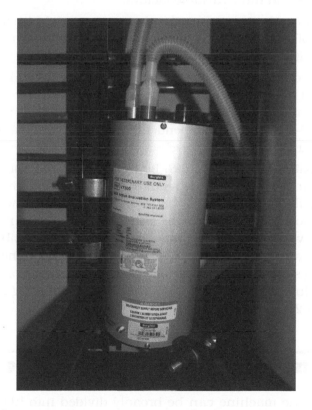

Fig. 2.10 An active scavenging pump that removes waste gas at 12 cubic feet/minute. On top, hoses from the anesthetic machine (left) and to the fume hood (right).

2.11 atmospheric pollution and occupational exposure

2.11.1 Atmospheric Pollution

Commonly used volatile anesthetic agents contribute to global warming through depletion of the ozone layer. This is compounded by their long lifetime in the atmosphere.

2.11.2 Occupational Exposure

Chronic exposure to trace concentrations of volatile anesthetics has been associated with a range of health problems, though conclusive evidence is often sparse due to inherent difficulties in studying complex systems and multivariable factors.

- Trace anesthetic exposure has been associated with increased rates of spontaneous abortion, mood disorders, headaches, infertility, cancer, and hepatic and renal disease.

- Eight-hour time-weighted average exposure limits have been defined for individual anesthetic agents, with a recommendation of a 50 parts per million (ppm) limit for isoflurane and 10–50 (country dependent) parts per million limit for halothane.

- Occupational exposure can be limited through some simple steps in the workplace: regular checking of equipment for leaks, flushing the anesthetic breathing system with fresh gas before disconnecting the animal at the end of anesthesia, filling vaporizers in a fume hood, avoiding chamber/mask induction and maintenance techniques, ensuring that a scavenging system is used routinely for all inhalational anesthetic procedures, and use of an effective air conditioning system (a rate of 20 air changes per hour is recommended).

2.12 anesthesia machine check out procedure

The anesthetic machine can be broadly divided into high and low-pressure systems. The high-pressure system includes everything between the inlet point of the carrier gas and the flowmeter, and the low-pressure system includes everything between the flowmeter and

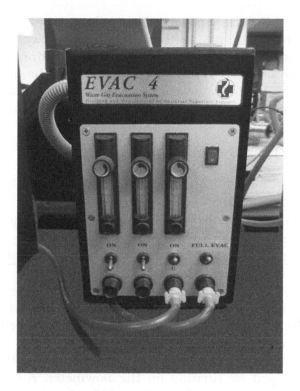

Fig. 2.11 A portable active scavenging pump.

the animal. If the anesthetic machine requires electricity to function, ensure it is connected to an electrical supply and switched on.

2.12.1 High-Pressure System

- Check the machine pressure gauges for pipeline gas; attached cylinders should read approximately 60 psi.
- Verify that there is an E cylinder available if planning to use pipeline gas.
- E cylinders should be at least half full (1000 psi) prior to starting an anesthetic procedure.

2.12.2 Low-Pressure System

- Flowmeter: Check that the flowmeter float moves through the full range of the sight tube when the oxygen supply is on.

- Vaporizer: Ensure that the concentration dial turns freely. Check that there is anesthetic in the vaporizer. Check that the vaporizer is correctly fitted to the machine. Check the vaporizer for leaks by temporarily occluding the common gas outlet with the flowmeter on; occlusion should result in a dip in the flowmeter float.

2.12.3 Breathing System

- Check that all breathing system components are present and are correctly arranged and fitted.
- System integrity: Perform a pressure leak test by occluding the patient end of the breathing system with the APL valve closed and flowmeter on until a pressure of 30 cm H_2O is reached on the breathing system pressure gauge. Once this pressure has been achieved, turn off the flowmeter and keep the breathing system occluded for 15 s, ensuring that the pressure does not drop. Pressure should be released from the system by opening the APL valve; this confirms the appropriate function of the APL valve. If a leak is identified, it can be quantified by turning on the flowmeter. A stable pressure with a flow rate of less than 100 mL/min is considered acceptable.
- Bain breathing systems should have the inner tube checked for leaks by occluding the inner tube at the patient end with the flowmeter on using the plunger from a syringe. A dip in the flowmeter float indicates a leak-free tube.
- Two bag test: Tests for patency of a circle system. Connect a reservoir bag to the patient end of the circle breathing system, turn on the flowmeter, and squeeze the reservoir bag on the machine with the APL valve closed. The unidirectional valves should open and close normally. Test the function of the APL valve by squeezing both reservoir bags simultaneously.

bibliography

Davey, A. J., A. Diba, and C. S. Ward. 2005. *Ward's Anaesthetic Equipment.* Elsevier Health Sciences Saunders.

Flecknell, P. A. and A. A. Thomas. 2015. Comparative anesthesia and analgesia of laboratory animals. In *Veterinary Anesthesia and Analgesia, the Fifth Edition of Lumb and Jones,* eds K. A. Grimm, L. A. Lamont, W. J. Tranquilli, S. A. Greens, and S. A. Robertson. 5th edition, 754–763. Ames, IA: Wiley.

Mosley, C. A. 2015. Anesthesia equipment. In *Veterinary Anesthesia and Analgesia, the Fifth Edition of Lumb and Jones,* eds K. A. Grimm, L. A. Lamont, W. J. Tranquilli, S. A. Greens, and S. A. Robertson. 5th edition, 23–85. Ames, IA: Wiley.

The Association of Anaesthetists of Great Britain and Ireland. http:// www.aagbi.org/ (accessed March 29, 2016).

anesthetic pharmacology and research-related anesthetic considerations

Robert D. Keegan

contents

3.1 preanesthetic drugs

3.1.1 Anticholinergics

- Decrease parasympathetic tone throughout the body and are thus associated with increases in heart rate and decreases in salivary secretions
- Are used to counteract bradycardia that is associated with administration of anesthetic drugs (opioids) or surgical procedures
- Are used to decrease oropharyngeal secretion that is associated with administration of anesthetic drugs

3.1.1.1 Atropine

- Naturally occurring belladonna extract having parasympatholytic properties
- Shorter duration of action compared with glycopyrrolate but more intense effect
- Effectiveness decreased in rats and rabbits (Sedgwick 1986) by an atropinease enzyme these animals produce

3.1.1.2 Glycopyrrolate

- Synthetic anticholinergic having a quaternary ammonium structure (Short 1987)
- Poorly lipid soluble due to the quaternary ammonium ring structure
- Longer duration of action compared to atropine but less intense effect
- Decreased volume of distribution and ability to cross the blood–brain and placental barrier

3.1.2 Tranquilizers

- Produce sedation and calm an animal prior to general anesthesia
- Potentiate the analgesic effects of opioids and sedative effects of barbiturates and propofol
- Decrease the dose of general anesthetic needed (MAC sparing)
- Promote smooth recovery from general anesthesia

3.1.2.1 Phenothiazines

- These offer predictable, dose-dependent sedation.
- The calming and sedative effects of these drugs, while predictable, are frequently inadequate to facilitate procedures; thus phenothiazines are commonly combined with an opioid.
- The dose-dependent blockage of α-1 adrenoceptors results in dose-dependent decreases in peripheral vascular resistance (PVR) and reductions in blood pressure (BP) (Ludders et al. 1983).
- They do not provide analgesia but will augment analgesia provided by opioids.

3.1.2.1.1 Acepromazine

- The most commonly used phenothiazine tranquilizer in veterinary anesthesia
- Provides synergistic sedative and analgesic effect when used in conjunction with opioids

3.1.2.2 Butyrophenones

- Offers variable, dose-dependent sedation.
- Blockage of α-1 adrenoceptors is less severe compared with phenothiazines, as is the reduction in blood pressure.

3.1.2.2.1 Droperidol

- Available as a separate drug and in combination with fentanyl as the neuroleptanalgesic drug Innovar-Vet

3.1.2.2.2 Fluanisone

- The butyrophenone component of the combination drug Hypnorm (Fentanyl/Fluanisone)

3.1.2.2.3 Azaperone

- Most commonly used in swine

3.1.2.3 Benzodiazepines

- Variable, dose-dependent sedation (less intense than that seen with phenothiazines).

- Act at endogenous benzodiazepine receptors to potentiate the action of γ-aminobutyric acid (GABA) within the central nervous system (CNS).
- Administration of benzodiazepines is generally not associated with hypotension.
- Useful for decreasing seizure activity within the brain.
- Promote relaxation of skeletal muscles.
- Provide no analgesia but will augment analgesia provided by opioids.
- The effects of benzodiazepines may be antagonized with the specific benzodiazepine antagonist flumazenil.

3.1.2.3.1 Diazepam

Non-water-soluble benzodiazepine that is prepared in solution with propylene glycol. Diazepam is generally not miscible with other agents owing to the propylene glycol vehicle.

3.1.2.3.2 Midazolam

Water-soluble benzodiazepine that is more potent than diazepam (Smith et al. 1991). The water-soluble nature of this drug permits mixing with other sedatives, analgesics, or anesthetics without causing precipitation.

3.1.2.3.3 Zolazepam

Water-soluble benzodiazepine that is available with tiletamine as a component of the drug Telazol (Zoletil).

3.1.2.4 α-2 adrenoceptor agonists

- Offer moderate to profound sedation (greater than that seen with the phenothiazines).
- Provide significant analgesia in most species.
- Act at endogenous α-2 receptors in the CNS, causing sedation and analgesia.
- Associated with significant cardiovascular effects such as hypertension, hypotension, and bradycardia (Lamont et al. 2001).
- Produce moderate to profound relaxation of skeletal muscles.
- When combined with opioids, α-2 agonists produce profound sedation and analgesia.

- The cardiovascular, sedative, and analgesic effects of α-2 agonists may be reversed with the α-2 antagonists yohimbine or atipamezole.

3.1.2.4.1 Xylazine

Offers moderate potency and moderate specificity for α-2 adrenoceptors.

3.1.2.4.2 Medetomidine

α-2 agonist having high potency and high specificity for α-2 adrenoceptors. It is a racemic mixture containing equal amounts of d- and l-isomers. Only the d-isomer has analgesic and sedative activity.

3.1.2.4.3 Dexmedetomidine

Offers high potency and high specificity for the α-2 adrenoceptor. It is composed of pure d-isomer and is thus twice as potent as medetomidine.

General comments concerning α-2 agonist sedative/analgesic.

- Produces profound sedation and analgesia, but after high dosages the righting reflex is not lost in mice.
- Produces intense peripheral vasoconstriction resulting in an initial period of hypertension and resultant vagally mediated bradycardia.
- Administration of α-2 antagonists (yohimbine or atipamezole) completely antagonizes both the sedative and analgesic properties of α-2 agonists. Excitement and or loss of analgesia may result.

3.1.3 Opioids (Please see "Pain Management," Chapter 7)

3.1.4 Neuroleptanalgesics

- Neuroleptanalgesic preparations contain an opioid mixed with a tranquilizer or sedative. The combination of the two drugs produces a synergistic analgesic and sedative effect.

3.1.4.1 Fentanyl-droperidol

- Available as the drug Innovar containing 0.4 mg fentanyl and 20 mg droperidol/mL.

- The fentanyl component may be antagonized with the opioid antagonist naloxone.

3.1.4.2 Fentanyl-fluanisone

- Available as the drug Hypnorm containing 0.315 mg/mL fentanyl and 10 mg/mL fluanisone.
- The fentanyl component may be antagonized with the opioid antagonist naloxone.

3.2 general anesthetics

3.2.1 Injectable Anesthetics

3.2.1.1 Dissociatives

3.2.1.1.1 Ketamine

- Cyclohexamine anesthetic derived from phencyclidine.
- Produces profound somatic analgesia, poor visceral analgesia.
- Acts as an antagonist at NMDA receptors and may help prevent "windup."
- Offers poor muscle relaxation; often administered with a tranquilizer or sedative to improve relaxation, that is, ketamine + xylazine, and so on.
- Produces cardiovascular stimulation (increases in heart rate, increases in systemic and pulmonary blood pressure, increases in cardiac output and myocardial work).
- Produces CNS stimulation to the point of seizures in some species; often administered with a tranquilizer or sedative to reduce CNS stimulation (White et al. 1982).

3.2.1.1.2 Tiletamine-zolazepam (Telazol®, Zoletil®)

- Combination drug containing tiletamine (a cyclohexamine anesthetic having greater potency than ketamine) and zolazepam (a water-soluble benzodiazepine).
- Offers similar cardiovascular and CNS properties to those associated with ketamine administration.
- Muscle relaxation is generally adequate owing to the relaxing effect of zolazepam.

3.2.2 Barbiturates

- Produce general anesthesia but are poor analgesics.
- Have a high pH and cause tissue irritation if injected extra-vascularly. Even though some barbiturates are administered intraperitoneally (IP), high concentrations may cause pain on injection (Berry 2015).

3.2.2.1 Pentobarbital

- A comparatively long-acting anesthetic that may be administered intravenously (IV) or IP.
- Narrow therapeutic index in that the anesthetic dose is close to the dose causing respiratory arrest.
- Potential for severe cardiovascular and respiratory depression.
- Poor analgesic.

3.2.2.2 Thiopental

- A short-acting anesthetic that is administered IV and produces rapid unconsciousness in many species.
- It is commonly used for rapid IV induction prior to inhalant anesthesia.
- It is irritating and can cause tissue reactions and sloughing if injected extravascularly.
- Accidental extravascular administration should be treated by injecting a 1:1 mixture of saline and 2% lidocaine equal to the volume of thiopental that was extravasated.
- Poor analgesic.
- Not currently available in the United States.

3.2.2.3 Methohexital

- Very short-acting barbiturate that produces rapid induction of anesthesia after IV administration in a wide range of species.
- Poor analgesic.
- Seizures during induction of anesthesia have been reported.
- Recovery from anesthesia may be accompanied by excitement and muscle tremors unless the animal is administered a pre-anesthetic tranquilizer or sedative.

3.2.3 Steroid Anesthetics

3.2.3.1 Alphaxalone/alphadolone (Saffan®)

- A mixture of two steroid anesthetic agents.

- Produces a smooth induction of anesthesia and may be administered IV, IM, or IP.

- The combination of the two steroids as Saffan® is not water soluble; thus, a polyoxyethylated castor oil preparation (Cremophor EL) is used as a solubilizing agent.

- Cremorphor EL has the potential to induce severe histamine release (in dogs); thus, Alphaxalone/alphadolone should not be used in dogs.

3.2.3.2 Alphaxalone (Figure 3.1)

- Alphaxalone has recently been released in a preparation that does not require alphadolone and Cremorphor EL (Ferre et al. 2006).

- May be administered IV, IM, or IP.

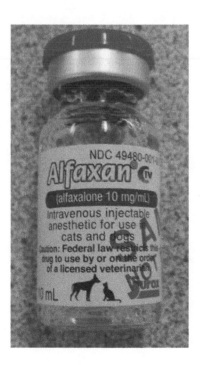

Fig. 3.1 Alfaxalone, a steroid injectable anesthetic.

3.2.4 Other Hypnotics

3.2.4.1 Propofol (Figure 3.2)

- A phenolic injectable anesthetic agent that is unrelated to barbiturate, cyclohexamine, or steroid anesthetics.
- Rapidly acting IV agent that is metabolized in the liver, lung, and spleen.
- Rapid metabolism and noncumulative effect make the drug attractive to administer by IV infusion.
- Anesthesia may be maintained by IV infusion and, once the infusion is discontinued, recovery occurs rapidly.
- Potent respiratory depressant; apnea during IV induction is common.
- The drug is a poor analgesic; thus, anesthetized animals must be provided with appropriate analgesia if painful surgical procedures are performed.
- Repeat administration of propofol, or administration of propofol via IV infusion (in cats), can cause oxidative damage to

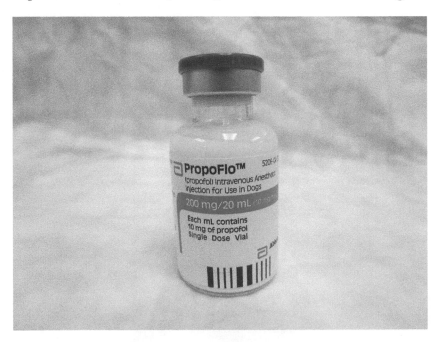

Fig. 3.2 Propofol (Abbot, Lake Forest, IL).

red blood cells, leading to methemoglobinemia (Andress et al. 1995).

3.2.4.2 Etomidate

- A short-acting imidazole anesthetic that produces minimal cardiovascular depression.
- Poor analgesic; thus, anesthetized animals must be provided with appropriate analgesia if painful surgical procedures are performed.
- May be administered IV or subcutaneously (SC).
- Associated with suppression of adrenocortical function following a single dose. Repeated doses or administration of etomidate by IV infusion results in further adrenocortical suppression (Fraser et al. 1984).

3.2.4.3 Tribromoethanol (Avertin®)

- A short-acting anesthetic used in rodents for producing short duration of anesthesia characterized by good muscle relaxation.
- Recovery from a single dose is rapid.
- The drug is irritating to the peritoneal lining and can cause peritoneal adhesions (Norris and Turner 1983).

3.2.4.4 Chloral hydrate

- A medium duration (1–2 h) anesthetic that has been administered IV to large domestic species such as horses and cattle and IP to rodents.
- The drug has minimal cardiovascular effects but is a potent respiratory depressant.
- Poor analgesic and has a low margin of safety when used for anesthesia.
- When administered IP to rodents, chloral hydrate has been reported to cause a high incidence of intestinal ileus.

3.2.4.5 Alpha-chloralose

- A long-acting injectable anesthetic that provides minimal analgesia (Flecknell 2009).

- Very long onset of action and will provide a very long duration of anesthesia (8–10 h).
- Minimal effect on the cardiovascular and respiratory systems; the activity of the autonomic nervous system is reported to be unaffected.
- Because recoveries are prolonged, alpha-chloralose is often used in rodent nonsurvival surgery.

3.2.4.6 Urethane

- A long-acting injectable anesthetic agent similar in duration to alpha-chloralose (8–10 h) but having a greater degree of analgesia.
- Minimal adverse effects on the cardiovascular and respiratory systems.
- Mutagenic and carcinogenic; as such, its use is limited to nonsurvival procedures.
- Laboratory personnel working with urethane must take adequate precautions to avoid inadvertent exposure.

3.2.5 Inhalant Anesthetics

Inhalant anesthetics are a group of anesthetics that are inhaled into the lung and gain entrance into the CNS. Most inhalant anesthetic agents are lipid-soluble agents and produce anesthesia by virtue of general CNS depression; they are poor analgesics, although they will potentiate the analgesic effect of opioids. Adverse health effects may occur if waste anesthetic agents are breathed by personnel; thus, it is imperative that administration of inhalant anesthetics occur only if adequate provision for capturing the waste anesthetic gases is provided. Solubility of the anesthetic in blood and tissues is a major determinant of the speed at which an inhalant anesthetic produces unconsciousness and the speed at which recovery from anesthesia occurs. Volatility of the inhalant anesthetic as measured by its vapor pressure as well as potency determine if the anesthetic may be easily administered via an open drop system or requires administration using an anesthetic vaporizer.

3.2.5.1 Diethyl ether

- Has relatively high blood and tissue solubility, translating to a relatively prolonged induction into and prolonged recovery from anesthesia

- Irritating to airways and may induce excessive salivation and breath holding during induction of anesthesia

- Explosive in concentrations that are necessary to maintain anesthesia (necessitates precautions by personnel and obviates the use of electrocautery)

3.2.5.2 Methoxyflurane (Figure 3.3)

- Potent inhalant anesthetic having a high blood and tissue solubility, translating to a relatively prolonged induction into and prolonged recovery from anesthesia.

- Low volatility and high potency make the drug attractive to administer in an open drop setting.

- Nonirritating to airways; thus, breath holding during induction of anesthesia does not usually occur.

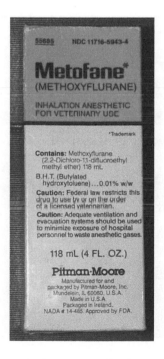

Fig. 3.3 Methoxyflurane, an inhalant anesthetic having high blood solubility. Owing to its high solubility, methoxyflurane is an attractive agent for use in open drop administration techniques.

3.2.5.3 Halothane (Figure 3.4)

- Inhalant anesthetic having a moderate blood solubility, translating to a relatively rapid induction into and recovery from anesthesia.

- High volatility and moderate potency enable administration in an open drop setting, although care must be taken to prevent to prevent inadvertent overdosage.

- Nonirritating to airways; thus, excessive salivation and breath holding does not usually occur.

- Sensitizes the myocardium to the action of endogenous and exogenous catecholamines to the point that cardiac arrhythmias commonly occur.

- Not currently available in the United States.

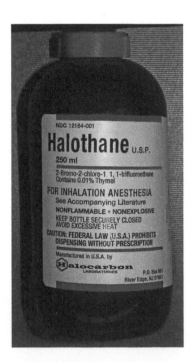

Fig. 3.4 Halothane, an inhalant anesthetic having moderate blood solubility. Owing to its moderate solubility, halothane is an attractive agent for use in open drop administration techniques.

3.2.5.4 Enflurane

- Inhalant anesthetic having a moderate blood and tissue solubility and giving a moderately rapid induction into and recovery from anesthesia.
- High volatility and moderate potency enable administration in an open drop setting, although care must be taken to prevent to prevent inadvertent overdosage.
- Induces CNS stimulation to the point of seizures.

3.2.5.5 Isoflurane (Figure 3.5)

- Inhalant anesthetic having a low blood and tissue solubility, translating to a rapid induction into and rapid recovery from anesthesia.
- High volatility and moderate potency enable administration in an open drop setting, although in light of the low blood solubility, care must be taken to prevent to prevent inadvertent overdosage.

Fig. 3.5 Isoflurane, an inhalant anesthetic having low blood solubility.

- Does not sensitize the myocardium to the action of endogenous and exogenous catecholamines to the degree seen with halothane; thus, cardiac arrhythmias are less frequent than with halothane.

3.2.5.6 Sevoflurane (Figure 3.6)

- Inhalant anesthetic having an extremely low blood and tissue solubility, translating to an extremely rapid induction into and recovery from anesthesia.
- High volatility and moderate potency enable administration in an open drop setting, although in light of the low blood solubility, care must be taken to prevent to prevent inadvertent overdosage.
- The low solubility may make it difficult to perform procedures on laboratory animals because the animals recover so quickly once administration is discontinued.

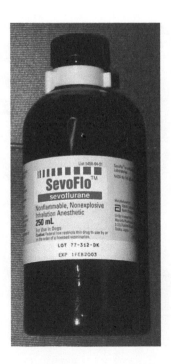

Fig. 3.6 Sevoflurane, an inhalant anesthetic having extremely low blood solubility.

3.2.5.7 Desflurane (Figure 3.7)

- Inhalant anesthetic having an extremely low blood and tissue solubility, translating to an extremely rapid induction into and recovery from anesthesia.

- Extreme volatility makes it impractical to administer desflurane in an open drop setting, and a precision desflurane vaporizer must be used.

3.2.5.8 Nitrous oxide

- Nitrous oxide has a very high vapor pressure and is thus present as a gas at room temperature. An anesthetic vaporizer is not necessary.

- Has low potency and it is not possible to induce or maintain anesthesia using nitrous oxide as the sole anesthetic.

- Has low blood and tissue solubility, giving the drug a rapid onset and rapid offset.

Fig. 3.7 Desflurane, an inhalant anesthetic having the lowest blood solubility of any available anesthetic.

- Often used in combination with a potent inhalant anesthetic such as isoflurane or sevoflurane to reduce the administered concentration of the potent anesthetic and the associated cardiopulmonary depression.
- Unlike other inhalant anesthetics, nitrous oxide possesses mild to moderate analgesic activity not related to generalized CNS depression.

3.2.6 Local Anesthetics (Please see "Pain Management," Chapter 7)

3.3 neuromuscular blocking agents

Neuromuscular blocking agents (NMBAs) are a group of drugs that act on the motor end plate of skeletal muscles to inhibit contraction in response to motor nerve activation (Keegan 2015). The NMBAs affect the contraction of skeletal muscle only and have no effect on smooth or cardiac muscle.

Indications: NMBAs are indicated whenever an animal's movement must be controlled or restricted during surgery. Examples include delicate surgical procedures wherein inadvertent patient movement would have adverse consequences and in animals who attempt to breathe against a mechanical ventilator.

Considerations

- Induce skeletal muscle paralysis; thus, the animal must be provided with positive pressure ventilation to ensure adequate exchange of oxygen and carbon dioxide.
- Have no anesthetic, sedative, or analgesic effects, and it is absolutely mandatory that they only be administered to animals that are adequately anesthetized and have adequate analgesia.
- Have activity at both pre- and postsynaptic acetylcholine receptors to block neuromuscular transmission. Most NMBAs are unstable at room temperature and must be stored under refrigeration.

3.3.1 Specific NMBAs

3.3.1.1 Succinlycholine

- The most rapidly acting NMBA available, having a dose-dependent onset time of approximately 60 s.

- Succinylcholine disrupts neuromuscular transmission by causing initial stimulation of postsynaptic, nicotinic acetylcholine receptors. This initial stimulation is uncoordinated and is visible grossly as body-wide muscle fasciculation.
- Not degraded as quickly as is acetylcholine and remains bound to the receptor, causing a persistent depolarization and resultant block of neuromuscular transmission and flaccid paralysis.
- Structurally similar to acetylcholine and has activity at cardiac autonomic ganglia. Changes in heart rate and rhythm can occur with administration of succinylcholine.
- Associated with an increase in serum potassium levels thought to be due to intracellular leakage of the ion.
- Although a specific antagonist is not available for a single administered dose of succinylcholine, the duration of action of the drug in most species is relatively short.

3.3.1.2 Pancuronium

- A long-acting NMBA having a duration of action of approximately 40–60 min and a dose-dependent onset time of approximately 5 min (Gleed and Jones 1982).
- Subsequent doses of pancuronium have a cumulative effect; thus, the drug is not often administered as a continuous infusion.
- Associated with a modest increase in heart rate that is due to a vagolytic property of the drug.
- Pancuronium neuromuscular blockade may be antagonized with the acetylcholinesterase antagonist drugs, neostigmine, or edrophonium.

3.3.1.3 Atracurium (Figure 3.8)

- An intermediate acting NMBA having a single-dose duration of action of approximately 25–30 min and an onset time of 5 min.
- Rapidly degraded by enzymatic hydrolysis and by spontaneous degradation at body temperature and pH. Accordingly, repeated doses of atracurium do not have a cumulative effect, making the drug attractive to administer by continuous rate infusion.

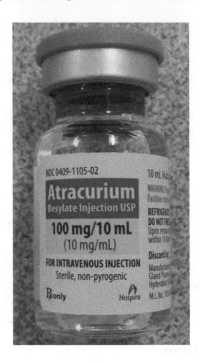

Fig. 3.8 Atracurium, a nondepolarizing neuromuscular blocking agent having a unique degradation pathway that involves spontaneous decomposition at body temperature and pH.

- Atracurium neuromuscular blockade may be antagonized with the acetylcholinesterase antagonist drugs neostigmine or edrophonium.

- Available as a racemic mixture of ten optical isomers. Administration is associated with histamine release and a potential decrease in blood pressure.

3.3.1.4 Cisatracurium

- An optical isomer of atracurium.

- The drug is approximately four times as potent and has a reduced potential for histamine release (Wastila et al. 1996).

3.3.1.5 Vecuronium

- An intermediate acting NMBA having a single-dose duration of action of approximately 25–30 min and an onset time of 5 min.

- The drug has remarkable cardiovascular stability and does not result in histamine release (Morris et al. 1983).
- Unlike most other NMBAs, vecuronium is supplied as a lyophilized powder that is reconstituted with sterile water prior to injection.
- The powder does not need refrigeration and, once reconstituted, the solution is stable for 24 h.

3.3.1.6 Mivacurium

- A rapid-acting, short-duration NMBA marketed for human tracheal intubation.
- Has the potential to release histamine, particularly if high dosages are administered rapidly as to facilitate tracheal intubation.
- The duration of action varies between species and has a duration of action of approximately 25 min in primates.
- The drug is more potent and is longer acting in dogs, with half of the primate dose lasting four times longer.

Rocuronium: Rocuronium is a derivative of vecuronium that has a more rapid onset of action and a similar duration of action in the dog.

Pipecuronium: Pipecuronium is a pancuronium derivative that retains the long duration of action of pancuronium (60 min) while being free from the vagolytic, tachycardiac effect.

3.3.1.7 Doxicurium

- A long-acting NMBA, similar in structure to atracurium and having a dose-dependent duration of action of 60 min (Martinez et al. 1998).
- Similar to cisatracurium, administration of doxicurium is not associated with an appreciable release of histamine.

3.3.1.8 Gantacurium

- A rapid-acting NMBA currently undergoing human clinical trials.
- Not stable in aqueous solution, and, similar to vecuronium, the drug is provided as a lyophilized powder that is reconstituted prior to administration.

- Has a unique degradation pathway that involves the binding to endogenous or exogenously administered cysteine and resultant inactivation of the drug (Lien 2002).

3.4 reversal agents

3.4.1 Cholinesterase Inhibitors

- The action of the NMBA may be reversed by administration of anticholinesterase drugs. These agents inhibit the normal rapid breakdown of endogenous acetylcholine, thus permitting the acetylcholine that is released at the nerve terminal in response to a motor nerve action potential to displace an NMBA from the motor end plate. Because these drugs inhibit the breakdown of acetylcholinesterase throughout the body, global vagal tone is increased and a decreased heart rate may result.

3.4.2 Neostigmine

- Produces a reversible inhibition of acetylcholinesterase by forming a carbamyl ester complex with the enzyme.
- Administration of neostigmine for reversal of an NMBA can result in excessive muscarinic effects such as bradycardia, miosis, salivation, and hyperperistalsis. Atropine is commonly administered to block these undesirable effects of neostigmine.

3.4.3 Edrophonium

- Produces a reversible inhibition of acetylcholinesterase by electrostatic attachment and hydrogen bonding to the enzyme.
- Shorter acting than neostigmine; administration is not usually associated with bradycardia, miosis, or salivation, thus atropine is not commonly administered.

3.4.4 Chelating and Adduction Agents

- The action of newer NMBAs such as rocuronium, vecuronium, and gantacurium may be reversed without the use of acetylcholinesterase-inhibiting drugs, thus avoiding potential muscarinic side effects.

3.4.4.1 Sugammadex

- A novel chelating agent that may be used as a reversal agent for the steroid ring NMBAs rocuronium and vecuronium (Naguib 2007; Adam et al. 2002).
- Preferentially binds to and removes the NMBA from the motor end plate, thus restoring neuromuscular transmission.

3.4.4.2 Cysteine

- The nonessential amino acid cysteine will bind to the ring structure of gantacurium and saturate a double bond within the molecule.
- Gantacurium is rendered inactive and normal neuromuscular transmission ensues.

---

references

Adam, J. M., D. J. Bennett, A. Bom, et al. 2002. Cyclodextrin-derived host molecules as reversal agents for the neuromuscular blocker rocuronium bromide: Synthesis and structure-activity relationships. *J. Med. Chem.* 45(9):1806–1816.

Andress, J. L., T. K. Day, and D. Day. 1995. The effects of consecutive day propofol anesthesia on feline red blood cells. *Vet. Surg.* 24(3):277–282.

Berry, S. H. 2015. Injectable anesthetics. In *Veterinary Anesthesia and Analgesia, the Fifth Edition of Lumb and Jones,* eds K. A. Grimm, L. A. Lamont, W. J. Tranquilli, S. A. Greens, and S. A. Robertson, 277–296. Ames, IA: Wiley-Blackwell.

Ferre, P. J., K. Pasloske, T. Whittem, M. G. Ranasinghe, Q. Li, and H. P. Lefebvre. 2006. Plasma pharmacokinetics of alfaxalone in dogs after an intravenous bolus of Alfaxan-CD RTU. *Vet. Anaesth. Analg.* 33(4):229–236.

Flecknell, P. 2009. Anaesthesia. In *Laboratory Animal Anaesthesia.* 3rd edition, 19–78. London: Elsevier.

Fraser, R., I. Watt, C. E. Gray, I. M. Ledingham, and A. F. Lever. 1984. The effect of etomidate on adrenocortical function in dogs before and during hemorrhagic shock. *Endocrinology* 115(6):2266–2270.

Gleed, R. D. and R. S. Jones. 1982. Observations on the neuromuscular blocking action of gallamine and pancuronium and their reversal by neostigmine. *Res. Vet. Sci.* 32(3):324–326.

Keegan, R. D. 2015. Muscle relaxants and neuromuscular blockade. In *Veterinary Anesthesia and Analgesia, the Fifth Edition of Lumb and Jones*, eds K. A. Grimm, L. A. Lamont, W. J. Tranquilli, S. A. Greens, and S. A. Robertson. 5th edition, 260–276. Ames, IA: Wiley.

Lamont, L. A., B. J. Bulmer, K. A. Grimm, W. J. Tranquilli, and D. D. Sisson. 2001. Cardiopulmonary evaluation of the use of medetomidine hydrochloride in cats. *Am. J. Vet. Res.* 62(11):1745–1749.

Lien, C. A. 2002. The pharmacology of GW280430A: A new nondepolarizing neuromuscular blocking agent. *Semin. Anesth. Periop. Med. Pain* 21(2):86–91.

Ludders, J. W., J. A. Reitan, R. Martucci, D. L. Fung, and E. P. Steffey. 1983. Blood pressure response to phenylephrine infusion in halothane-anesthetized dogs given acetylpromazine maleate. *Am. J. Vet. Res.* 44(6):996–999.

Martinez, E. A., A. A. Wooldridge, S. M. Hartsfield, and K. L. Mealey. 1998. Neuromuscular effects of doxacurium chloride in isoflurane-anesthetized dogs. *Vet. Surg.* 27(3):279–283.

Morris, R. B., M. K. Cahalan, R. D. Miller, P. L. Wilkinson, A. L. Quasha, and S. L. Robinson. 1983. The cardiovascular effects of vecuronium (ORG NC45) and pancuronium in patients undergoing coronary artery bypass grafting. *Anesthesiology* 58(5):438–440.

Naguib, M. 2007. Sugammadex: Another milestone in clinical neuromuscular pharmacology. *Anesth. Analg.* 104(3):575–581.

Norris, M. L. and W. D. Turner. 1983. An evaluation of tribromoethanol (TBE) as an anaesthetic agent in the Mongolian gerbil (*Meriones unguiculatus*). *Lab. Anim.* 17(4):324–329.

Sedgwick, C. J. 1986. Anesthesia for rabbits. *Vet. Clin. North Am. Food Anim. Pract.* 2(3):731–736.

Short, C. E. 1987. Anticholinergics. In *Principles and Practice of Veterinary Anesthesia*, ed. C. E. Short, 8–15. Baltimore, MD: Williams and Wilkins.

Smith, A. C., J. L. Zellner, F. G. Spinale, and M. M. Swindle. 1991. Sedative and cardiovascular effects of midazolam in swine. *Lab. Anim. Sci.* 41(2):157–161.

Wastila, W. B., R. B. Maehr, G. L. Turner, D. A. Hill, and J. J. Savarese. 1996. Comparative pharmacology of cisatracurium (51W89), atracurium, and five isomers in cats. *Anesthesiology* 85(1):169–177.

White, P. F., W. L. Way, and A. J. Trevor. 1982. Ketamine: Its pharmacology and therapeutic uses. *Anesthesiology* 56(2):119–136.

management of anesthesia

Jennifer C. Smith

contents

4.1 introduction

Management of anesthesia in any species requires proper planning, implementation, and resources to ensure a humane and successful outcome. When working with rodents, it becomes especially critical to consider species-specific physiology when planning the appropriate anesthetic and analgesic methodology to be used. The management of any anesthetic event can be organized into three basic periods: pre-anesthetic, perianesthetic, and postanesthetic. This chapter covers all three of these unique periods with specific details relating to the proper planning and implementation of rodent anesthesia (Table 4.1).

4.2 preanesthetic period

One of the most important, and often abbreviated or omitted, components of proper anesthesia is the time spent preparing for the anesthetic event. The preanesthetic period represents the largest amount of time and effort for the laboratory animal professional, as proper assessment, planning, and implementation are essential to obtaining a humane and successful anesthetic outcome. The selection of anesthetic agents for use in rodents is essential; however, so also is the formulation of the plan for utilizing these agents. Thus, preparation and planning become

Table 4.1: CHAPTER TERMINOLOGY

Anesthetic event
The anesthetic event describes the process of planning and delivering anesthesia for a specific purpose (i.e., surgery, restraint).

Anesthetic/analgesic plan
An anesthetic plan defines the protocol of anesthesia and analgesia to be followed during the entire anesthetic event. This plan is specific to the animals that have been evaluated and should be updated each time an assessment or modification is made.

Preanesthetic period
The preanesthetic period describes the time frame prior to the delivery of any anesthetic. This period represents the largest portion of effort due to the proper planning and implementation necessary for a successful outcome.

Perianesthetic period
The perianesthetic period describes the time frame that the animal is undergoing anesthesia. This includes the use of preanesthetics and the induction, maintenance, and monitoring of anesthesia.

Postanesthetic period
The postanesthetic period describes the time frame after anesthesia has concluded and the animal is recovering. The focus of this period is proper supportive care, pain assessment, and analgesia support to assure the animal returns to its preanesthetic physiologic state.

---

the hallmark features of the preanesthetic phase and can best be summarized into three categories: preparation of the patient, preparation of the equipment, and planning for the anesthetic event.

4.2.1 Preparation of the Patient

Every anesthetic event has some inherent risk associated with it due to the pharmacologic nature of the majority of anesthetic and analgesic agents used. These agents often cause a dose-related physiologic depression on the cardiopulmonary system and can result in unexpected morbidity and mortality. The most important element in reducing this risk is assuring that the animals undergoing anesthesia appear healthy and free of any disease state prior to the start of anesthesia (Flecknell 2009). Performing a basic but necessary evaluation of all rodents receiving anesthesia is an essential first step in assuring a successful anesthetic outcome.

- Species physiology and body condition can impact the response to anesthetic and analgesic agents.
  - Rodents have a high ratio of surface area to body weight.
  - Rodents are prone to anesthesia-induced hypothermia.
  - Their small body size makes intubation and IV access difficult.
- Basic physiologic data for the rat and the mouse have been described previously (Danneman et al. 2013; Sharp and Villano 2013) and are summarized in Tables 4.2 and 4.3.
- Obtaining high-quality rodents with known health status will reduce the overall risk of anesthetic morbidity and mortality.

4.2.1.1 Patient assessment

This is an essential first step in assuring a successful anesthesia outcome. For rodents, this assessment may be conducted on the individual animal or on the cage/study group level. The latter is more common in a high-throughput research setting utilizing rodents. These assessments made prior to the anesthetic event become the foundation for comparison when the patient enters the postanesthetic period. The following should be taken into account when performing a thorough patient assessment prior to the delivery of any anesthetic:

- An animal/cage/group assessment should be performed in the animal's holding room.

Table 4.2: MISCELLANEOUS BIOLOGICAL PARAMETERS OF THE MOUSE

Parameter	Typical Value
Diploid chromosome number	40
Life span	2–3 years
Adult body weight	20–40 g
Body temperature	36.5°C–38.0°C (97.5°F–100.4°F)
Metabolic rate	180–505 kcal/kg/day
Food intake	12–18 g/100 g body weight/day
Water intake	15 mL/100 g body weight/day
Respiratory rate	80–230 breaths/min
Heart rate	500–600 beats/min

Source: Republished with permission of CRC Press Taylor & Francis Group, from Danneman, P. J. et al., *The Laboratory Mouse*, 2nd edition, CRC Press, Boca Raton; London; New York, 2013, page 9, Table 1.1 copyright 2013; permission conveyed through Copyright Clearance Center, Inc.

Table 4.3: BASIC BIOLOGICAL PARAMETERS IN RATS

Parameter	Value
Lifespan (year)	2.5–3.5
Adult male body weight (BW) (g)[a]	450–520
Adult female BW (g)[a]	250–300
Body temperature	35.9°C–37.5°C (96.6°F–99.5°F)
O_2 consumption (mL/m²/g BW)	0.84
Body surface area (cm²)	10.5 (BW in g)$^{2/3}$
Food intake (g/100 g BW/d)	5–6
Water intake (mL/100 g BW/d)	10–12
GI transit time (h)	12–24
Urine volume (mL/100 g BW/d)	5.5
Total body water (mL)[a]/250 g BW	167
Intracellular fluid (mL)[a]/250 g BW	92.8
Extracellular fluid (mL)[a]/250 g BW	74.2

Source: Republished with permission of CRC Press Taylor & Francis Group, from Sharp, P., and J. S. Villano, *The Laboratory Rat*, 2nd edition, CRC Press, Boca Raton, 2013, page 7, Table 2 copyright 2013; permission conveyed through Copyright Clearance Center, Inc.
[a] Body weight (BW) will vary with stock or strain.

- The following parameters should be assessed:
  - Activity level
  - Posture and mobility
  - Hydration status

- Nutritional status
- Food/water intake
- Feces/urine output
- Body condition scoring (BCS) should be performed:
  - BCS provides a measurable and recordable assessment of the overall physiologic state of the animal.
  - BCS is a proven method for assessing the general well-being state of the animal prior to the anesthetic event (Hickman and Swan 2010; Ullman-Cullere and Foltz 1999) (Figure 4.1).
  - A BCS score can be assigned to each animal/cage/group without disruption or added stress.
- Normative species behaviors should be observed and taken into account when planning for anesthesia and analgesia (Danneman et al. 2013; Sharp and Villano 2013).
- Behavioral and well-being assessments relating to nest building and enrichment usage should also be noted (Hess et al. 2008).
- The body weight should be obtained to accurately calculate dosage of anesthetic agents.
- Animals that appear ill or those that do not exhibit normal rodent behavior should be evaluated by the veterinary staff prior to the administration of any anesthetics.
- All pertinent information should be included in the anesthetic and study records for future reference.

4.2.1.2 Patient acclimation

It is commonly accepted that a period of time should be made available to allow any research animal to acclimate to its surroundings prior to the commencement of research activities. This practice is particularly important for rodent species, as they are commonly exposed to stress due to shipment and rehousing. Providing an acclimation period helps to minimize any nonexperimental variables, while also ensuring humane animal care and use (Conour et al. 2006).

- A quarantine/acclimation period as mandated by the Institutional Animal Care and Use Committee (IACUC) is a reasonable approach to assuring that some type of acclimation period is included in the research protocol utilizing anesthesia.
- In rodents, the general accepted acclimation period prior to any procedure is at least 48–72 h.

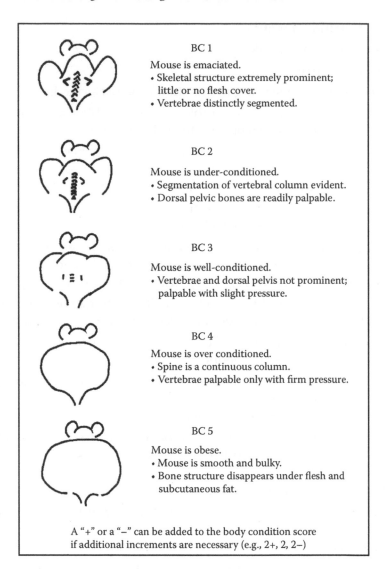

Fig. 4.1 Body condition scoring. (Reprinted with permission from AALAS. Ullman-Cullere, M.H. and C.J. Foltz, *Comp. Med.*, 49, 319–323, 1999.)

4.2.1.3 Patient fasting

The physiology of most research species necessitates the use of fasting prior to the administration of anesthesia. Rodent species, however, are not prone to the deleterious effects of not fasting prior to administration of anesthesia and as such are not routinely fasted (Struck et al. 2011).

- Rodents have a high metabolic rate, requiring a constant food and water supply.
- Rodents do not possess the anatomical structures for vomiting during induction, obviating the need for fasting prior to anesthesia.

4.2.2 Preparation of the Anesthetic Equipment

The following items should be assessed prior to the start of any anesthetic event to assure they are present and functioning properly:

- Anesthetic machine: Check to make sure all of the components are present, clean, and functional, including the vaporizer, carrier gas source and flow meter, and waste anesthetic gas scavenging device (Flecknell 2016).
- Anesthetic delivery tools: The induction chamber, face mask, tubing, and breathing circuit should be appropriate for species and size; intubation materials should include restraint device, light source, and tubing if applicable (Rivera et al. 2005; Molthen 2006).
- Anesthesia monitoring equipment: Select the device(s) most appropriate for the species and procedure, including capnography, pulse oximetry, electrocardiography (ECG), blood pressure, and temperature recording.
- Supplemental heat source: A wide variety of electronic and nonelectronic devices can be used to combat anesthesia-induced hypothermia as summarized by Hankenson (2014).
- Anesthetic/analgesic drugs: Check to make sure all inject able drugs are present and within expiratory usage dates. Include Drug Enforcement Administration (DEA) forms and storage requirements as applicable per location.
- Ancillary equipment: This should include infusion pumps for delivery of fluids or constant rate infusions; syringes, needles and catheters for parenteral injections; restraint devices; and a weight scale appropriate for species size.

4.2.3 Anesthetic and Analgesic Planning

Once the necessary assessments have been made, the anesthetic and analgesic plan can be formulated and implemented.

- The following questions should be taken into consideration when the anesthetic event is being planned:
  - What is the purpose of the anesthesia? (i.e., surgical procedure, restraint)
  - How long is the intended duration?
  - Where will the procedure take place?
  - What special equipment will be needed?
  - What are the postoperative and analgesic needs?
- The anesthetic plan should include all aspects of the anesthetic event and take into account the species-specific and physiologic assessments that have been made.
- A thorough search of the current literature should be performed to assure that all aspects of anesthesia and analgesia have been addressed.
- The plan should include agents for both anesthesia and analgesia.
- The plans should be consistent with what has been approved in the corresponding IACUC protocol.
- The plan should be communicated with all members of the husbandry, veterinary and research teams.

Examples of completed anesthetic plans are represented in Tables 4.4 and 4.5.

Table 4.4: SAMPLE ANESTHETIC PLAN (MOUSE)

Species: Mouse
Stock or Strain: BALB/c
Sex: Female
Age: 10 weeks
Weight: 20 g
Body condition score: 3
Procedure needing anesthesia: Restraint needed for replacement ear tagging
Acclimation time: 48 h
Fasting time: None
Equipment needed: Anesthesia machine, circulating water blanket
Premed: None
Induction: Isoflurane in chamber
Maintenance: Isoflurane by facemask as needed
Recovery: In recovery cage on circulating water blanket until standing
Analgesic plan: Buprenorphine (0.05–0.1 mg/kg) given PRN
Postop plan: Assess every 15 min until recovered

Table 4.5: SAMPLE ANESTHETIC PLAN (RAT)

Species: Rat

Stock or strain: SD

Age: 12 weeks

Sex: Male

Weight: 300 g

Body condition score: 3

Procedure needing anesthesia: Surgical anesthesia for femoral venous catheter insertion

Acclimation time: 48 h

Fasting time: None

Equipment needed: Anesthetic machine, circulating water blanket, pulse oximitry

Premedication: None

Induction: Ketamine (80 mg/kg) and Xylazine (10 mg/kg) given IP

Maintenance: Isoflurane by facemask if needed

Recovery: In recovery cage on circulating water blanket until standing

Analgesic plan: Buprenorphine (0.01–0.005 mg/kg, SC) given at time of surgery; additional doses PRN; Meloxicam (1 mg/kg, SC) given postoperatively

Postop plan: Assess every 15 min until recovered; assess pain every 8–12 h thereafter; give analgesics PRN

4.3 perianesthetic period

Once an anesthetic event has been properly planned and initiated, the perianesthetic period can begin. This period contains the actual anesthetic delivery and is often represented by vigilant monitoring of both the patient and the surroundings. The perianesthesia period includes preoperative/procedural anesthesia, maintenance, and immediate recovery.

- During this time period, the anesthetic plan should be followed as directed; however, emergent physiological changes must also be addressed as they arise.

- Asking a series of physiologic questions and utilizing the anesthetic record to document this information is the best way to assure a humane and successful anesthetic event.

- The balance between what has been planned for and what is actually observed is the hallmark of good anesthetic management and care.

4.3.1 Preanesthesia

This period of the anesthetic event includes the proper handling, restraint, and premedication of rodents prior to the induction of anesthesia.

- Proper restraint and handling. Working with laboratory rodents often necessitates the use of some type of restraint in order to effectively perform the task while assuring a humane and successful outcome. Restraint methods are commonly classified into different categories (manual, mechanical, and chemical).

4.3.1.1 Manual restraint methods

- Includes the handling of rodents in order to restrain them for a specific procedure.
- Most of these common techniques described can be further adapted for use with one-person (Figure 4.2) or two-person handling.

4.3.1.2 Mechanical restraint methods

- Include the use of any device that can safely and effectively restrain rodents.

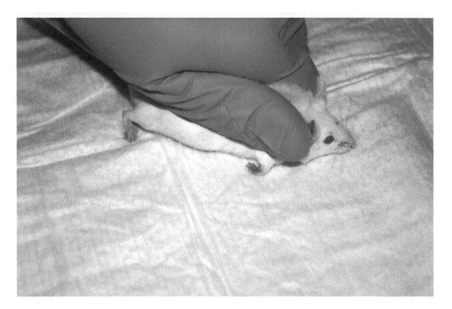

Fig. 4.2 Proper initial restraint of the mouse.

- These devices are often purchased but can also be made and/ or modified for specific use.
- Examples of mechanical restraint methods include
  - DecapiCones
  - Bowman restrainer
  - Clear plastic flat bottom restrainer
  - Solid brass restrainer
- A thorough video review of common rodent restraints suitable for anesthesia is described by Machholz et al. (2012).

4.3.1.3 Chemical restraint methods

- Include the use of chemical agents to effectively restrain rodents for anesthetic delivery.
- Most rodents can be safely handled without the need for premedication before anesthesia induction; however, the following agents can be used where chemical restraint is warranted:
  - Acepromizine
  - Ketamine
  - Xylazine
  - Medetomidine
  - Dexmedetomidine

4.3.2 Anesthetic Induction and Maintenance Period

Once the preanesthesia period has been completed, the induction and maintenance of anesthesia can occur. The hallmark of this time period is monitoring of the patient and the mitigation of anesthetic-related morbidity and mortality.

- Many monitoring devices have been microscaled for specific use with rodents.
- Monitoring equipment made for larger species can also be adapted for rodent use.
- Current trends in human anesthesia utilize monitoring equipment with mathematical algorithms for consistent monitoring and recording.

- The following should be continually monitored during rodent anesthesia:
  - Anesthetic depth
  - Temperature
  - Cardiopulmonary status
    - Mucus membrane color (i.e., ear, tail, paw color)
    - Heart rate/rhythm
    - Respiratory rate/pattern
  - Ancillary monitoring equipment includes:
    - Pulse oximitry
    - Blood pressure
    - Capnography
    - ECG

4.4 postanesthetic period

The postanesthetic period encompasses the period of time after the anesthesia has concluded and the animal is recovering and returning to its preanesthetic physiologic state. Once the basic recovery has occurred, there are several physiological processes that should be evaluated and addressed.

4.4.1 Patient Assessment

Just as the evaluation of the animal/cage/group of rodents was essential to the preanesthetic period, these observations are equally important in the postanesthetic period. When possible, a cage side monitoring card should be used (Figure 4.3) to document assessment.

- The following parameters should be observed and assessed during the postanesthesia period:
  - Activity level and mobility
  - Posture
  - Porphyrin staining

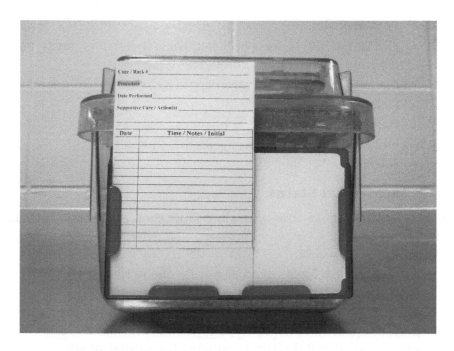

Fig. 4.3 A cage side monitoring card. (Courtesy of Cholawat Pacharinsak.)

- Hydration status
- Food/water intake
- Feces/urine output

4.4.2 Assessment of Pain

The aforementioned assessments are an important factor in determining the postoperative physiologic state of the animal; however, they are not necessarily specific identifiers for pain. Additionally, some postoperative behaviors may be a result of anesthesia and not always a specific indicator of pain (Wright-Williams et al. 2007).

- Rodents are traditionally prey species, making them less likely to demonstrate overt pain behaviors as some larger species do.
- Other methods of pain assessment should be utilized for rodents, including

- Pain scoring
- Evaluation of other species-specific behaviors such as grooming and nest building
- Metabolic markers indicative of pain (i.e., corticosterone levels)
- Body weight
- Body temperature

(Please also refer to "Pain Management," Chapter 7)

4.4.3 Nutritional Status

Several factors make the nutrition status of the postanesthetic patient important:

- Rodents have a high metabolism rate due to their high energy requirements and their limited fat storage.
- Rodents reach a hypermetabolic state during and after surgery, frequently causing hyperglycemia and an increase in tissue protein catabolism to support the wound repair.
- The delivery of anesthesia alone can cause moderate changes in these physiologic parameters, regardless of surgical model.
- A postanesthesia diet plan should be discussed prior to anesthesia and should include
  - Acclimation to diet for several days before the anesthetic event
  - Provision of a postanesthesia diet that is highly flavored and palatable
  - Making food and water easily accessible
  - Recording food/water intake as well as feces/urine production
  - Considering adding analgesia to postanesthesia diet

4.4.4 Suggested Postanesthetic (Postoperative) Care

- Soft bedding (Alpha—dri) (Figure 4.4)
- Soft food on the floor
- Water access on the floor (i.e., clear H_2O gel) (Figure 4.5)
- Nesting material (i.e., Enviro Dri) (Figure 4.6)

Fig. 4.4 Alpha-Dri (Shephered, Milford, NJ). (Courtesy of Cholawat Pacharinsak.)

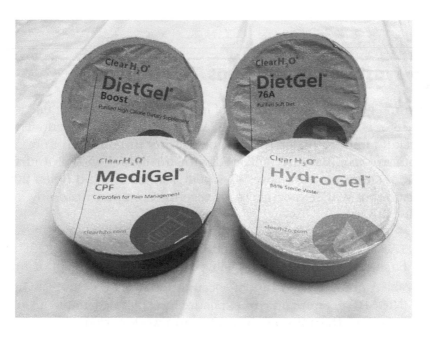

Fig. 4.5 Clear H$_2$O (Westbrook, ME). (Courtesy of Cholawat Pacharinsak.)

Fig. 4.6 Enviro Dri (Shepherd, Milford, NJ). (Courtesy of Cholawat Pacharinsak.)

4.5 waste anesthetic gas

A common occupational hazard of inhalational anesthesia use is the release of waste anesthetic gas (WAG) into the environment. This is especially important for the rodent user as inhalational anesthetic agents are traditionally delivered with an induction box and face-mask, resulting in excessive WAG pollution (Smith and Bolon 2002). Mitigation of WAG exposure is an essential component to any anesthetic plan.

- The Occupational Safety and Health Administration (OSHA) has yet to establish permissible exposure levels for isoflurane.

- The National Institute of Occupational Safety and Health (NIOSH) recommends an exposure level of halogenated agents to be less than 2 ppm over a 60-min work period.

- IACUC-mandated WAG exposure limits are a reasonable approach to assuring the safety of patients and personnel.

In keeping with the current exposure recommendations, the following WAG mitigation practices should be incorporated into anesthesia management when using halogenated anesthetic agents in rodents:

- Halogenated agents should be used in a calibrated precision vaporizer whenever possible.
- If the open drop method is used, careful consideration and planning for patient and operator safety is essential (Taylor and Mook 2009).
- Careful consideration of equipment and work practices should be employed, including
  - Utilizing proper vaporizer filling devices to prevent leakage
  - Utilizing proper storage to prevent breakage and spills of bottled halogenated agents
  - Release of induction chamber gases prior to opening
  - Shutting off the vaporizer and carrier gas prior to induction chamber opening
  - Placing the induction chamber in a chemical fume hood
  - Assuring a tight-fitting seal around the animal and facemask
  - Incorporating endotracheal intubation practices
  - Properly checking for leaks prior to the start of anesthesia
  - Properly cleaning the induction box and tubing after use
  - Recovering animals in a well-ventilated and dedicated space
- WAG scavenging devices should be employed whenever halogenated agents are in use:
  - Activated charcoal canisters can be effective if properly used and maintained (Smith and Bolon 2003).
  - Active scavenging systems can be purchased or created using the existing facility exhaust system.
  - If available, chemical fume hoods can be used to reduce WAG exposure.
  - Respirators can be worn when working with halogenated agents, but the user must comply with the occupational safety and health program.

- Monitoring of WAG exposure is an essential component to any occupational safety and health program and should include
  - Real-time monitoring with a spectrophotometer device
  - Badge monitoring to establish a time-weighted average (TWA) of exposure

references

Conour, L. A., K. A. Murray, and M. J. Brown. 2006. Preparation of animals for research: Issues to consider for rodents and rabbits. *ILAR J.* 47(4):283–293.

Danneman, P. J., M. A. Suckow, and C. F. Brayton. 2013. Important biological features. In *The Laboratory Mouse*, 2nd edition, 1–21. eds P. J. Danneman, M. A. Suckow and C. F. Brayton. Boca Raton; London; New York: CRC Press.

Flecknell, P. 2009. Preparing for anaesthesia. In *Laboratory Animal Anaesthesia*, 3rd edition. London: Elsevier.

Flecknell, P. 2016. Basic principles of anaesthesia. In *Laboratory Animal Anaesthesia*, 4th edition. London: Elsevier.

Hankenson, F. C. 2014. General approaches for critical care. In *Critical Care Management for Laboratory Mice and Rats*, 1–25. Boca Raton, FL: CRC Press.

Hess, S. E., S. Rohr, B. D. Dufour, B. N. Gaskill, E. A. Pajor, and J. P. Garner. 2008. Home improvement: C57BL/6J mice given more naturalistic nesting materials build better nests. *J. Am. Assoc. Lab. Anim. Sci.* 47(6):25–31.

Hickman, D. L. and M. Swan. 2010. Use of a body condition score technique to assess health status in a rat model of polycystic kidney disease. *J. Am. Assoc. Lab. Anim. Sci.* 49(2):155–159.

Machholz, E., G. Mulder, C. Ruiz, B. F. Corning, and K. R. Pritchett-Corning. 2012. Manual restraint and common compound administration routes in mice and rats. *J. Vis. Exp.* (67). pii: 2771. doi:10.3791/2771.

Molthen, R. C. 2006. A simple, inexpensive, and effective light: Carrying laryngoscopic blade for orotracheal intubation of rats. *J. Am. Assoc. Lab. Anim. Sci.* 45(1):88–93.

Rivera, B., S. Miller, E. Brown, and R. Price. 2005. A novel method for endotracheal intubation of mice and rats used in imaging studies. *Contemp. Top. Lab. Anim. Sci.* 44(2):52–55.

Sharp, P. and J. S. Villano. 2013. Important biological features. In *The Laboratory Rat*, 2nd edition, 1–30. eds P. Sharp and J. S. Villano. Boca Raton, FL: CRC Press.

Smith, J. C. and B. Bolon. 2002. Atmospheric waste isoflurane concentrations using conventional equipment and rat anesthesia protocols. *Contemp. Top. Lab. Anim. Sci.* 41(2):10–17.

Smith, J. C. and B. Bolon. 2003. Comparison of three commercially available activated charcoal canisters for passive scavenging of waste isoflurane during conventional rodent anesthesia. *Contemp. Top. Lab. Anim. Sci.* 42(2):10–15.

Struck, M. B., K. A. Andrutis, H. E. Ramirez, and A. H. Battles. 2011. Effect of a short-term fast on ketamine-xylazine anesthesia in rats. *J. Am. Assoc. Lab. Anim. Sci.* 50(3):344–348.

Taylor, D. K. and D. M. Mook. 2009. Isoflurane waste anesthetic gas concentrations associated with the open-drop method. *J. Am. Assoc. Lab. Anim. Sci.* 48(1):61–64.

Ullman-Cullere, M. H. and C. J. Foltz. 1999. Body condition scoring: A rapid and accurate method for assessing health status in mice. *Lab. Anim. Sci.* 49(3):319–323.

Wright-Williams, S. L., J. P. Courade, C. A. Richardson, J. V. Roughan, and P. A. Flecknell. 2007. Effects of vasectomy surgery and meloxicam treatment on faecal corticosterone levels and behaviour in two strains of laboratory mouse. *Pain* 130(1–2):108–118.

anesthesia monitoring

Andre Shih

contents

General anesthesia in rodents is often induced and maintained with inhalational, injectable, or combination anesthetics. General anesthesia and/or anesthetics cause severe cardiovascular depression, vasodilation, and decrease in organ perfusion; therefore, anesthesia should always be accompanied by some cardiopulmonary monitoring. The degree of monitoring will vary from simple observation to multiparameter invasive data observation (Table 5.1), depending on the procedure being performed. Goals for anesthesia monitoring are to detect physiological changes in time to start therapeutic intervention (Gargiulo et al. 2012).

Table 5.1: SIMPLE OBSERVATION OF ANESTHETIC PLANE

Parameters	Light Plane	Surgical Plane	Deep Plane
Spontaneous movement	Present	Absent	Absent
Jaw tone	Present	Absent	Absent
Paw withdrawal reflex	Present	Absent	Absent
Mucous membrane (mm) color	Pink	Pink	Pale/gray
Respiratory pattern	Rapid	Rapid/shallow	Slow/gasping

Source: The table modified from Adams, S. and C. Pacharinsak. *Curr. Protoc. Mouse Biol.,* 5(1), 51–63, 2015 and based on the author's experience.

5.1 multiparameter

Multiparameter (cardiopulmonary) monitoring can be challenging in rodents.

- Blood-pressure cuffs, electrocardiogram (ECG), pulse oximeter probes, and other disposable units were not designed to fit such small animals.

- Small patient size makes placing arterial or venous catheter difficult (Gargiulo et al. 2012).

- Basal metabolic rate, heart rate, and respiratory rate are normally higher in rodents than in humans. Human anesthesia monitoring equipment is not designed to work in these conditions and can give inaccurate readings. Equipment designed for the veterinary market might perform better but it is still liable to error.

- The choice of monitoring depends on various considerations:

- There are a range of procedure variables (species, size, health status, duration of procedure, etc.).

- The operator's expertise in using and familiarity with the monitoring device still represent an important factor in obtaining reliable results. No monitoring device can improve an outcome by itself (Adams and Pacharinsak 2015; Davis 2008).

- To positively affect the outcome, the obtained values should be accurate. The results should also be of clinical relevance to trigger a therapeutic intervention, and the therapeutic intervention made must be proved to improve the outcome. If the information obtained is inaccurate or interpreted incorrectly, the resultant intervention will be unlikely to improve the patient's condition and could potentially be detrimental.

5.2 commonly used monitoring equipment during anesthesia

5.2.1 Electrocardiogram (ECG)

- An ECG shows the electrical activity of the heart during the perioperative period. As such, the ability to understand and interpret the information is important for all anesthesia-related personnel (Farraj et al. 2011).

- Surface leads are designed to detect low voltage signals generated by the heart during a normal cardiac cycle. In rodents, the standard ECG leads are too big and cumbersome to be used (Ho et al. 2011). An example ECG is shown in Figure 5.1.

- A small-gauge needle (29G or smaller) can be placed in the subcutaneous region in place of the ECG alligator clips. Once the generated signal enters the monitor, it is conditioned, amplified, and displayed on a panel. Another option is to use surface contact adhesive pads with or without the need for ECG leads (Figure 5.2) (Farraj et al. 2011).

- This will allow ECG configuration to read in I, II, and II.

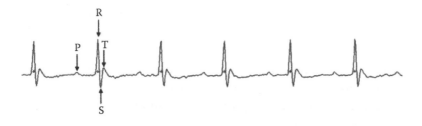

Fig. 5.1 An example of ECG in a mouse. Permitted from John Wiley & Sons, Inc. Copyright © 2013 all rights reserved. David Ho et al. Heart rate and electrocardiography monitoring in mice. *Curr. Protoc. Mouse Biol.* 2011. 1. 123–139.

(a)

(b)

Fig. 5.2 Example of noninvasive ECG measurement using footplate electrodes in adult (a) and neonatal (b) animals. (Courtesy of Mouse Specifics, Inc., Framingham, MA.)

- ECG monitoring during anesthesia is very useful to diagnose arrhythmias and decreases in HR, but it *cannot* be used to diagnose chamber enlargement.
- ECG does not monitor cardiac contraction, and a patient in cardiac arrest can still have an ECG rhythm. As such, information obtained during anesthesia monitoring is limited to electrical *rate* and *rhythm*.
- Some of the causes of bradyarrhythmias (like atrial ventricular block or sinus bradycardia) are severe hypoxia, increase in vagal tone, difficult intubation, severe hypothermia, deep plane of anesthesia, iatrogenic drugs, or primary cardiac disease.

- Some of the causes of tachyarrhythmias (like ventricular tachycardia or sinus tachycardia) are hypovolemia, noxious stimulation, light plane of anesthesia, anaphylactic reaction, iatrogenic drugs, or primary cardiac disease.

5.2.2 Blood Pressure

- Blood pressure (BP) is the pressure exerted by the circulating blood against the vessel walls.
- It gives an estimation of cardiac contraction (cardiac output [CO]) and tissue perfusion.
- BP is influenced not only by CO but also by systemic vascular resistance (SVR). A patient can have an adequate CO with a low SVR yielding a low BP. On the other hand, a normotensive patient does not necessarily have a normal CO (Shih et al. 2011).
- During each heartbeat, pressure varies between maximum (systolic arterial blood pressure [SAP]) and minimum (diastolic arterial blood pressure [DAP]). SAP can be an indication of the stroke volume (SV) and arterial compliance. DAP is more dependent on how the SVR is performing. The mean arterial pressure (MAP) is an average of both values and can be an estimation of average driving tissue perfusion pressure.
- A rodent's intrinsic fast pulse rate and decreased arterial pulsatile signal, combined, lowers the accuracy of the BP monitoring device.
- Measurements are less accurate during hypotension, and when small arteries are used. All noninvasive BP monitors can give erroneous values if there is excessive patient motion or if the cuff is not the correct size. Cuff width should be approximately 40% of the circumference of the limb. Narrow and wide cuff widths provide erroneously high and low pressure measurements, respectively.

5.2.2.1 Blood pressure measurement

5.2.2.1.1 Noninvasive BP

- *Doppler flow detection* is used for the audible monitoring of blood flow. It is based on the Doppler principle: The frequency of transmitted sound waves is altered when reflected off moving red blood cells. As cuff pressure slowly decreases and pulsatile arterial blood flow begins, an audible Doppler signal is detected corresponding to systolic arterial pressure (Schaefer et al. 2005). The Doppler flow detector uses a probe placed as close as possible to blood flow source (an artery or the heart). There are three types of probes: (1) human adult, (2) pediatric, and (3) pencil. The pediatric probe works relatively well in the base of the tail of rodents. The Doppler can also be used over the eye globe as a "flow" pulse rate detector. This is particularly useful in extremely hypotensive animals and during cardiopulmonary arrest resuscitation.

- *The oscillometric technique* uses a deflating cuff and a monitor to detect oscillations of the encircled artery. As cuff pressure slowly decreases, intermittent blood flow is detected at the SAP point of maximum oscillations on the manometer, roughly corresponding to the MAP. DAP is read when oscillations are no longer detected and is the least accurate of the three measurements. Most of the time, oscillometric BP monitors detect SAP and MAP and calculate DAP by the equation MAP = (SAP + 2 × DAP)/3. Sites for indirect BP measurement include the legs and tail (Figure 5.3).

5.2.2.1.2 Invasive BP

- A gold standard and would provide beat-by-beat SAP, MAP, and DAP information. Its accuracy can decrease if the pressure system is damped. One must make sure that the extension set and pressure transducer are made of a noncompliant material, that there are no air bubbles in the set system, and that there is no excessive length of tubing between the arterial catheter and the transducer. Direct arterial catheterization is used less frequently than indirect because it is technically difficult. Tail and femoral arterial catheterization is feasible in rodents. When performing arterial catheterization, aseptic techniques must be followed to avoid infection and bacteremia (Gargiulo et al. 2012; Schaefer et al. 2005).

- *Hypotension* is a common occurrence during general anesthesia (Zuurbier et al. 2002).
  - It can be an indication of pharmacologically induced vasodilation (a drop in SVR), decreased contractility, decreased stroke volume, sepsis, hemorrhage, or hypovolemic shock (Shih et al. 2011).
  - If hypotension is persistent, it can lead to a decrease in tissue perfusion, precipitate anaerobic metabolism,

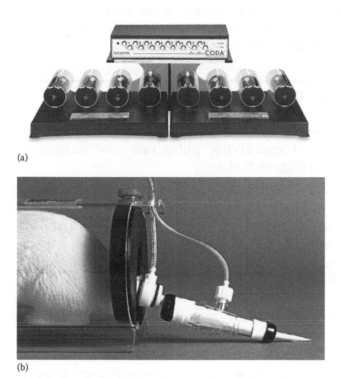

(a)

(b)

Fig. 5.3 (a) CODA® tail-cuff blood pressure set-up system. It measures systolic, diastolic, mean BP, heart rate, tail blood volume, and blood flow. (b) CODA® tail cuff. (Courtesy of Kent Scientific Corporation, Torrington, CT.)

metabolic acidosis, and ischemic organ damage (kidney, liver, and brain injury).

- *Therapy* includes decreasing the pharmacological agents that are causing the vasodilation (e.g., inhalant anesthetics), restoring circulating blood volume (fluid bolus or transfusion), or administering positive inotropes (dopamine, dobutamine, vasopressin, or norepinephrine infusion).

- *Hypertension* can be an indicator of increase in catecholamine release due to pain or an inadequate depth of anesthesia.
  - It is usually treated with sedatives and analgesics.
  - Persistent hypertension can be pharmacologically induced or due to hypermetabolism disease.

- *Therapy* includes removing the underlying causes. Drugs like nitroglycerine and nitroprusside can decrease blood pressure by decreasing SVR.

5.3 pulse oximeter

- Pulse oximeter is a simple, inexpensive and noninvasive method of monitoring pulse rate and oxygen saturation (%SpO_2) (Figure 5.4).
  - It provides continuous information regarding oxygenation.
  - It is based on the Lambert–Beer law, which relates to the absorption of light by hemoglobin (Hb) to the concentration of available oxygen. Oxygenated Hb absorbs light in the infrared spectrum (920–960 nm), whereas deoxygenated Hb absorbs light in the red spectrum (640–660 nm). Analysis of the differential absorption spectra of oxygenated and deoxygenated hemoglobin

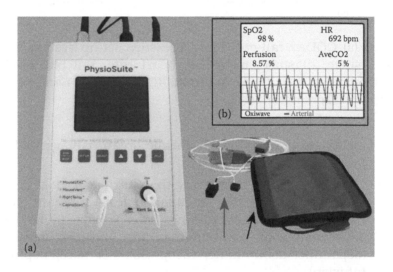

Fig. 5.4 (a) PhysioSuite unit for pulse oximeter (Mouse STAT), ventilator (MouseVent), body temperature (RightTempo), and capnometer (CapnoScan). Probes for pulse oximeter (red arrow) and heating pad (black arrow). (b) Reading example of PhysioSuite (Kent Scientific Corporation). SpO_2 is 70%–100%, heart rate (HR) range is up to 900 bpm, $AveCO_2$ range is 0%–16% or 0–120 mmHg.

by the microprocessor in the monitor provides a value for oxygen saturation. Pulse oximeter requires adequate plethysmographic pulsations to allow it to distinguish arterial light absorption. It is inaccurate in the presence of decreased BP, decreased pulse pressure, vasoconstriction, pigment (dark color skin), excessive patient movement, ambient lighting, carboxyhemoglobin (COHb—falsely high reading), methhemoglobin (MetHb—falsely low reading).

- In general, pulse oximeter tends to be most accurate within the normal range (SpO_2 95%–100%).

- Sites for placement of pulse oximeter probes in rodents are the ears, tongue, buccal mucosa, paw, vulva, prepuce, tail, and thigh. A reflectance pulse oximeter probe can be used in the esophagus, rectum, or cloaca. In rats, based on the author's experience, the best site is the base of the tail or thigh (not the ears due to excessive compression of the aural vasculature) (Gargiulo et al. 2012).

- O_2 saturation is just part of the oxygen delivery (DO_2) equation, and a normal O_2 saturation does not always mean normal blood perfusion or normal DO_2 (Santos et al. 2015).

$$DO_2 = CaO_2 \times CO$$

$$CaO_2 \, (ml/dl) = \left(1.36 \times \lfloor Hb \rfloor \times \%SpO_2\right) + \left(PaO_2 \times 0.003\right)$$

$$CO = HR \times SV$$

where:
DO_2 = O_2 delivery
CaO_2 = O_2 content
CO = cardiac output
HR = heart rate
SV = stroke volume
Hb = hemoglobin
PaO_2 = partial pressure of O_2 in arterial blood

5.4 capnography

- Capnography is the measurement of respiratory CO_2 concentration during the respiratory cycle. It can be one of the most useful pieces of equipment in rodent anesthesia.

- End-tidal CO_2 ($ETCO_2$) is the maximum CO_2 concentration that is measured at the end of expiration. $ETCO_2$ can be measured continuously and in real time. $ETCO_2$ should be ~ 35–45 mmHg (Thal and Plesnila 2007).

- Capnography/$ETCO_2$
  - Estimates partial pressure of CO_2 in arterial blood ($PaCO_2$)
  - Confirms correct endotracheal intubation
  - Assesses patient ventilatory and circulatory status
  - Alerts anesthetist during apnea
  - Ensures anesthetic circuit integrity
  - Helps in cardiovascular monitoring during CPR
  - Evaluates cardiorespiratory and anesthetic machine functions

- The waveform is the only normal shape for the capnograph. The wave can be divided into four phases. Phase I is the mid part of the inspiration. It is represented by the E–A distance. In this phase, carbon dioxide should be zero, which means that the patient is breathing in fresh gas. Phase II is the start of expiration (BC). Its fast rising line represents the mixing between the dead space gas, which is full of fresh gas from the last inspiration, with the alveolar carbon dioxide. Phase III is the rest of the expiration, representing exhaled alveolar gas (CD). This phase is also called the plateau portion of expiration. The angle α should be between 90° and 110°. D is the highest point of the wave and represents the $ETCO_2$ (Schaefer et al. 2005). The $ETCO_2$ is the closest estimation of $PaCO_2$. Phase IV is the start of inspiration (DE). The decline on CO_2 concentration should go down to zero and the angle β should be close to 90°.

- Lung disease increases differences between $ETCO_2$ and $PaCO_2$.

- The most obvious sampling error is a system leak. Another sampling problem is related to the sampling rate and the minute ventilation of the patient. When used in small patients, high sampling rates (i.e., 250 mL/min) can result in entrainment of room air and lower $ETCO_2$ values. In human pediatric patients, the sampling rate is set at 50 mL/min, and this rate is still too high in many small patients (Thal and Plesnila 2007).

- In rodents, due to their rapid respiratory rate and small tidal volume capnograph waveform, details are hard to interpret. It can be difficult to use waveform to detect abnormalities:

  - *Hypercapnia* (high $ETCO_2$) is the same as hypoventilation. Possible causes are deep plane of anesthesia, respiratory depression due to opioid use, obesity, or patients' positions.

  - *Hypocapnia* (low $ETCO_2$) is the same as hyperventilation. Possible causes are decrease in CO, decrease in BP, pulmonary thromboembolism (PTE), or decrease in pulmonary perfusion (cardiac arrest).

  - *Airway obstruction*: Waveform will be almost zero.

- $ETCO_2$ cannot replace $PaCO_2$. Cardiorespiratory changes as well as certain conditions may increase the $ETCO_2$ and $PaCO_2$ differences.

- Capnography is a great tool for the clinician during anesthesia; however, judicious interpretation must be made, and therefore measurement of $PaCO_2$ may be necessary to make a more accurate assessment of ventilator status.

5.5 body temperature

- Because rodents have a high surface area compared to body size (volume), they easily lose body heat. In addition, any anesthetic drugs cause peripheral vasodilation and suppress normal thermoregulatory mechanisms leading to hypothermia once anesthetized.

- Hypothermia has been correlated with several morbidities, for example, delayed healing, hypercoagulability, pain, and delayed metabolism.

- Body temperature measurement is a standard of care during all anesthetic procedures. Perioperative rectal and esophageal temperature can be easily, inexpensively and noninvasively recorded throughout the anesthetic period.

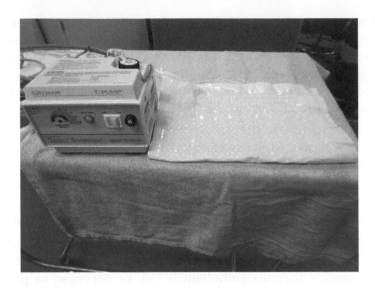

Fig. 5.5 Water circulating heating pad. (Courtesy of Cholawat Pacharinsak.)

Water circulating heating pads are highly recommended, and they should be used from the beginning of anesthetic induction until full recovery (Figure 5.5) (Zuurbier et al. 2002).

5.6 arterial blood gas

Arterial blood gas (ABG [Table 5.2]) assesses patient oxygenation, acid–base status, and adequacy of ventilation that is from carotid or femoral artery. Venous blood gas (VBG) samples are more easily obtained than are arterial samples in rodents, for example, from the tail vein (Thal and Plesnila 2007).

- Most portable handheld point of care ABG analyzers do not require a lot of blood to run. If repeated measurements are needed, operator should be conscientious and avoid removing more then 10% of animal total blood volume.
- Respiratory alkalosis (high pH with low PCO_2 and HCO_3) usually indicates hyperventilation and can be due to chronic hypoxia, pain, or a light plane of anesthesia.
- Respiratory acidosis (low pH and high PCO_2 and HCO_3) usually indicates hypoventilation and can be due to respiratory disease, neurologic disease, or a deep plane of anesthesia.

Table 5.2: BLOOD GAS VALUES IN RATS

	Arterial (100%O_2)	Venous
pH	7.59 ± 0.17	7.38 ± 0.12
PO2 (mmHg)	346.08 ± 149.88	37.17 ± 13.93
PCO2 (mmHg)	22.5 ± 17.77	43.42 ± 15.94
HCO3 (mmol/l)	19.05 ± 3.76	24.53 ± 3.33
BE	−1.02 ± 3.55	−1.29 ± 2.87
% Saturation	100 ± 0.00	70.60 ± 17.33
Hct %	35.17 ± 4.72	38.58 ± 3.53

Source: Reprinted from Son K.H. et al., *Clin. Hemorheol. Microcirc.*, 44(1), 27–33, 2010. Copyright 2010, with permission from IOS Press.

Table 5.3: ANESTHETIC MONITORING AND RECOVERY RECORDS SHOULD BE PERFORMED FOR EACH ANIMAL DURING ANESTHESIA

Protocol #_____

PI_____ Surgeon_____

Animal ID _____ Species_____

Body weight _____g

Date_____ Start _____ End _____

Procedure_____

_____

Anesthetics_____

Fluid_____

Time (min)	00	15	30	45
Isoflurane %				
Withdrawal reflex	YN	YN	YN	YN
Resp.Pattern	FastSlow	FastSlow	FastSlow	FastSlow
mm color	PinkBlue	PinkBlue	PinkBlue	PinkBlue

Recovery				
Time				
Resp. Pattern	FastSlow	FastSlow	FastSlow	FastSlow
mm color	PinkBlue	PinkBlue	PinkBlue	PinkBlue

- Metabolic acidosis (low pH and low PCO_2 and HCO_2) usually indicates increase in cations and can be due to lactic acidosis, diabetes ketoacidosis, renal uremic acidosis, hyperchloremic acidosis, or iatrogenic administration of acids.

Anesthetic monitoring and recovery records should be performed for each animal during anesthesia (Table 5.3).

references

Adams, S. and C. Pacharinsak. 2015. Mouse anesthesia and analgesia. *Curr. Protoc. Mouse Biol.* 5 (1): 51–63.

Davis, J. A. 2008. Mouse and rat anesthesia and analgesia. *Curr. Protoc. Neurosci.* Appendix 4: Appendix 4B.

Farraj, A. K., M. S. Hazari, and W. E. Cascio. 2011. The utility of the small rodent electrocardiogram in toxicology. *Toxicol. Sci.* 121 (1): 11–30.

Gargiulo, S., A. Greco, M. Gramanzini, et al. 2012. Mice anesthesia, analgesia, and care, Part II: Anesthetic considerations in pre-clinical imaging studies. *Ilar J.* 53 (1): E70–E81.

Ho, D., X. Zhao, S. Gao, C. Hong, D. E. Vatner, and S. F. Vatner. 2011. Heart rate and electrocardiography monitoring in mice. *Curr. Protoc. Mouse Biol.* 1: 123–139.

Santos, A., L. Fernandez-Friera, M. Villalba, et al. 2015. Cardiovascular imaging: What have we learned from animal models? *Front. Pharmacol.* 6: 227.

Schaefer, A., G. P. Meyer, B. Brand, D. Hilfiker-Kleiner, H. Drexler, and G. Klein. 2005. Effects of anesthesia on diastolic function in mice assessed by echocardiography. *Echocardiography* 22 (8): 665–670.

Shih, A., S. Giguere, A. Vigani, R. Shih, N. Thuramalla, and C. Bandt. 2011. Determination of cardiac output by ultrasound velocity dilution in normovolemia and hypovolemia in dogs. *Vet. Anaesth. Analg.* 38 (4): 279–285.

Son, K. H., C. H. Lim, E. J. Song, K. Sun, H. S. Son, and S. H. Lee. 2010. Inter-species hemorheologic differences in arterial and venous blood. *Clin. Hemorheol. Microcirc.* 44 (1): 27–33.

Thal, S. C. and N. Plesnila. 2007. Non-invasive intraoperative monitoring of blood pressure and arterial pCO_2 during surgical anesthesia in mice. *J. Neurosci. Methods* 159 (2): 261–267.

Zuurbier, C. J., V. M. Emons, and C. Ince. 2002. Hemodynamics of anesthetized ventilated mouse models: Aspects of anesthetics, fluid support, and strain. *Am. J. Physiol. Heart Circ. Physiol.* 282 (6): H2099–H2105.

special techniques and species

Cholawat Pacharinsak

With Janlee Jensen

contents

6.1 imaging anesthesia

6.1.1 General Considerations

- Anesthetics can impact multiple body systems, so the researcher should ensure that the anesthetic used does not interfere with that under study. In particular, cerebral metabolism and neurotransmitters are often affected. Carrier gases (i.e., air, O_2, etc.) may also affect results.

- The animal's physiological response to anesthetics must be considered. For example, CO_2 can affect cerebral blood flow (CBF). This is especially important in positron emission tomography (PET) scanning and functional magnetic resonance (MR) (Figure 6.1).

- Too many variables exist to cover in this chapter, so the researcher must refer to the current literature when planning each study. Alstrup and Smith (2013) provide a good review article for PET scanning of the brain, while Hanusch et al. (2007), McConville (2011), and Driehuys et al. (2008) discuss anesthesia for MR.

- Imaging can be very quick or prolonged. Sometimes heavy sedation, rather than full anesthesia, is adequate. Sometimes no anesthesia at all is needed, especially for ultrasound, but also for some PET or functional magnetic resonance imaging (fMRI) exams. Even if anesthesia is required for the imaging, it may not be required for other aspects of the study such as tracer uptake. However, animals should be trained and adapted to the restraint so that stress does not alter results.

- If general anesthesia is needed, an inhalant or continuous rate infusion (CRI) of an injectable anesthetic is generally best. However, the researcher must remember that anesthetic response varies not only with species but also with strain, gender, fed or fasted state, time of day, and many other factors. Therefore, the anesthetic must be tailored to

 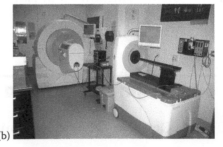

(a)   (b)

Fig. 6.1 (a) 3T MRI scanner. (b) MicroPet and MicroPET/CT scanners. (Courtesy of Dr. Timothy Doyle, Stanford University.)

the individual animal as well as to the imaging target. For example, fasting for 6 h may be necessary to empty the stomach for the imaging.

- When beginning a new project, consider including a few extra animals to allow the anesthesia and support to be optimized for both the patient and the research. When these factors have been optimized, they should be standardized so that all animals are handled and anesthetized in the same way.

- When an animal will be catheterized in an imaging system, use a splint on the catheterized area so that it doesn't kink.

- Body temperature should be kept at the desired temperature (~ 98.6°F–100.4°F), both for maintenance of the animal's anesthetic plan and because functional imaging results can vary with temperature. Self-regulating thermal devices based on rectal or esophageal temperature are best, but avoid devices that may inflate or move and decrease image resolution, such as flexible warm-water pads. The animal, including its tail, can be wrapped in plastic wrap or a blanket to reduce heat loss.

- Intubation and ventilation are often desirable to avoid hypoventilation and atelectasis.

- Several rodent intubation kits (i.e., Kent Scientific, Harvard apparatus, etc.) are commercially available. As with any rodent anesthesia, circuit dead space must be minimized. Ideally, $PaCO_2$ or $ETCO_2$ should be monitored.

- Image data can be gated with the cardiac or respiratory cycle if needed, but sometimes neuromuscular blocking agents (NMBAs) are preferred for higher resolution imaging.

- For MRI or spectroscopy, ferromagnetic components must be avoided. MR-compatible thermometers, warming devices, pulse oximetry sensors, electrocardiographic leads, ventilators, and anesthetic machines are available. Compatible pneumatic sensors are also available to alert operators to the cessation of respiration in nonventilated animals. If compatible anesthetic machines are not available, long tubing can be used to connect the animal to an anesthetic machine kept outside the MR room, or an injectable anesthetic can be used. Ensure that the animal does not have any ferromagnetic implants. For consistent positioning, a special molded bed can be used. Some beds can be designed to include physiological monitoring and support devices. Small Animal Instruments,

Talavera Science, Biopac Systems, and other companies can provide integrated equipment for small-animal imaging.

6.2 long-term anesthesia

- For purposes of this text, *long-term anesthesia* is defined as that lasting from 3 h up to several days.

- Prophylactic antibiotics are generally recommended for survival procedures or for nonsurvival procedures lasting over 8–12 h. The antibiotic must be given such that it is present in the blood at the time of the initial surgical incision.

  - Rats and mice: cefazolin (30–40 mg/kg, IP) can be given initially 30 min prior to procedure, then continued IP, IM, IV or IO every 6 h.

  - Hamsters, gerbils, and guinea pigs: enrofloxacin (5–10 mg/ kg, IP, IM, IV, IO) can be given once daily.

- Fluid therapy (5–10 ml/kg/h) is usually necessary. For non-survival procedures lasting under 8 h, intermittent SC or IP fluid therapy may be acceptable, but hypotension should prompt IV fluid administration. Any balanced crystalloid fluids (0.9% NaCl, LRS etc.) can be used.

- For procedures over 48 h, add water-soluble vitamins (injectable B complex solution at 1 ml/l of fluids for all, plus daily 30 mg veterinary injectable vitamin C IM for guinea pigs). The addition of amino acids is recommended for procedures over 48–72 h. Most animals require about 5 g protein per 100 kcal. Amino acid solutions over 6% should be given into only large (central) veins.

- Potassium (K) and magnesium (Mg) supplementation may also be required during these longer procedures. In general, K can be provided at 20 mEq/L if the fluid is given at a maintenance daily rate, but it is best to measure the patient's electrolytes and supply as necessary. Patients should never be given over 0.2 mEq/kg/h of potassium without intensive monitoring, because high levels of potassium may cause cardiac arrest.

- Every 4 h, the patient's position, generally between left lateral recumbency, right lateral recumbency, and sternal recumbency, should be changed if the procedure permits it. Maintain the patient on a padded surface and inspect pressure points (generally the ventral thorax, elbows, carpi, knees, and tarsi) for any signs of skin compromise. Gently massage large muscle groups and move joints through their normal range of motion. Reposition pulse oximeter sensors every 2–4 h.

- Lubricate the eyes every 2–4 h with a bland ophthalmic ointment, and tape them closed with a gentle medical tape (one that does not strip hair off when it is removed).

- Ensure the patient is clean, gently washing the fur coat with a baby wipe if necessary. Any surgery should be done aseptically if the patient is to survive or if the procedure is to last over 8–12 h. The surgical site, if left open, should be kept moist and sterile with sterile saline–soaked gauze sponges or laparotomy sponges. Any percutaneous catheter sites should be covered with a sterile pad and bandaged; these sites should be cleaned with chlorhexidine and rebandaged every 24 h. Signs of infection (such as redness, discharge, local swelling of the catheter site) should prompt removal of the catheter, culture if indicated by the length of the procedure, and replacement at another site.

- The animal's body temperature must be supported with a warm water circulating pad or a forced-air warmer.

- If anesthesia is to last > 4 h, the bladder must be gently expressed every 4 h. Avoid pressing too hard, which could rupture the bladder in small patients.

- In most cases, anesthesia is best performed with inhalant anesthesia, often plus a CRI of opioids (fentanyl, etc.), propofol, midazolam, lidocaine, and so on.

- In all cases, even those not using NMBAs, respiration must be closely monitored and a ventilator available in case positive pressure ventilation is needed. Tracheostomy can be used in nonsurvival procedures. Peroral endotracheal intubation is required for survival procedures, but it can also be used as a refinement in nonsurvival procedures. Commercially available rodent intubation packs can be very helpful.

- Change tie position every 4 h and gently cleanse the mouth and swab the pharynx with a chlorhexidine solution manufactured for oral use.

- Wrap the tongue with saline-moistened gauze every 2–4 hours.

- The tube should be suctioned gently every 8 h for less than 5 s to remove secretions and other debris.

- Airway humidification (Figure 6.2) should be provided to prevent airway dessication.

- The animal may breathe unenriched air (approximately 21% O_2) for some procedures, but often an increased fraction of inspired O_2 (FiO_2) is required. Because long-term high oxygen levels may lead to oxygen toxicity, and because oxygen can worsen atelectasis, added oxygen should ideally be used at the lowest level possible. For example, if oxygen is required, begin at about 0.3 (30%) FiO_2, and increase as necessary. If possible, a $FiO_2 > 0.5$ (50%) should not be used.

- Ventilators are typically used in a mandatory mode, which means that the ventilator completely controls respiration. The ideal parameters to avoid ventilator-induced lung injury are still controversial, but typical starting recommendations (van den Brink et al. 2013) for ventilator settings in otherwise healthy patients include

  - Tidal volume (TV) should be 5–10 ml/kg.

  - Peak inspiratory pressure should be 5–15 cm H_2O.

  - Respiratory rates are generally set at 50–80 (guinea pigs) breaths per minute (BPM) and 80–170 BPM (smaller mammals).

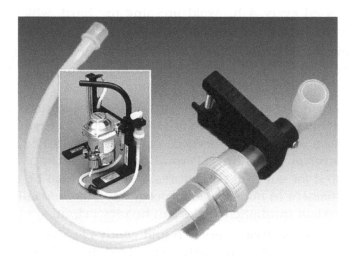

Fig. 6.2 Airway humidification. (Courtesy of Ryan Sullivan, Product Research and Development Patterson Scientific (division of Patterson Vet) 1160 Chess Drive, Suite 9 Foster City, CA 94404.)

- Once a ventilator is initiated, anesthetic inhalation (i.e., isoflurane) should be decreased accordingly.

- In some ventilators (i.e., PhyisoSuite™ from Kent Scientific), the user has only to position the animal and the machine will automatically run the ventilator (weight range 3–500 g).

- Inhalants are probably the best choice for prolonged anesthesia. Like injectable anesthetic infusions, they may be titrated down to the lowest effective dose. Opioids and other analgesics can be given for painful procedures to reduce the amount of inhalant needed.

- For prolonged anesthesia, repeated doses of short-acting anesthetics may be used, but this approach can be risky because individual responses to anesthesia vary.

  - Propofol formulated in oil may cause lipemia. This may interfere with experimental measurement of many blood or serum parameters. "Propofol infusion syndrome" has been reported in humans due to prolonged used (Rajda et al. 2008). This syndrome includes the failure of multiple organs and usually results in death, and the pathophysiology is not known.

  - Etomidate is another very rapidly acting drug, which allows fast awakening upon termination of the infusion. Be aware that etomidate will suppress the adrenocortical response.

  - Remifentanil, an opioid, is rapidly degraded by tissue and plasma esterases, so animals awaken quickly after the infusion stops. Fentanyl, sufentanil, and alfentanil are still relatively short acting, but their effects will be prolonged with long-term infusion.

  - Thiobutabarbituate (Inactin), urethane, and α-chloralose are all non-pharmaceutical-grade drugs. Note that α-chloralose is very slow to take effect, requiring 15 min after IV administration and longer after IP administration. In addition, it may cause myoclonic jerks. Both chloralose and urethane have minimal depression on the cardiovascular and respiratory systems at light levels of anesthesia, but depression of both may occur at surgical doses. Additional analgesia or anesthesia may be needed for major surgery with both of these drugs. Urethane is mutagenic and carcinogenic and can cause toxicity to the viscera and mesenteric blood vessels when given IP. Urethane is also hazardous to staff. Applicable

United States Department of Agriculture (USDA), Office of Laboratory Animal Welfare (OLAW), Institutional Animal Care and Use Committee (IACUC), and institutional environmental health and safety laws and regulations should be followed when using these drugs. All of these drugs are recommended only for nonsurvival surgery.

- For commonly used long-term anesthetic drugs, see Tables 6.1 through 6.3.

Table 6.1: LONG TERM ANESTHESIA IN RATS

Drug (mg/kg)

Pentobarbital (55) IP, then Remifentanyl 0.4 µg/kg/min IV.

Remifentanyl (0.2–0.4) IV if response (Ismaiel et al. 2012).

Inhalant to effect + fentanyl 12.5 µg/kg/h IV.

Ketamine (37) + xylazine (7) IM, then up to 12 h of nonsurgical anesthesia with 40–50 µl/min of mixture of 0.5 ml of 100 mg/ml ketamine + 0.08 ml of 20 mg/ ml xylazine in 2.5% dextrose and 0.45% saline (Simpson 1997).

Propofol 25 mg/kg/h (Tung et al. 2001); consider adding an opioid to decrease propofol dose.

Dexmedetomidine 30–180 µg/kg/h (Tung et al. 2008).

Dexmedetomidine 25 µg/kg IM + Telazol (20) IP, then pentobarbital (10) hourly IV for up to 5 h (Ferrari et al. 2012).

Pentobarbital (50) IP, followed by 6–30 mg/kg/h IV, or 10 mg/kg boluses IP every hour or as needed.

Thiobutabarbital (Inactin) 80 mg/kg IV or IP lasts 1–4 h, but gives reliable surgical anesthesia only if given IV.

α-chloralose (55–65) IP provides 8–10 h of light anesthesia.

Urethane 1 g/kg IP provides 3–4 h of anesthesia. Higher doses (up to 1500 mg/ kg IP) have been used but with increased mortality.

Urethane (250–400) IP, then 30 min later α-chloralose (35–40) IP, produced anesthesia for up to 6 h (Hughes et al. 1982).

Table 6.2: LONG-TERM ANESTHESIA IN MICE

Drug (mg/kg)

Propofol (26) IV then 120–150 mg/kg/h IV. Add opioids to decrease propofol dose.

Inhalant to effect + remifentanyl 40–160 µg/kg/h IV (Cabanero et al. 2009).

Isoflurane + fentanyl (0.025) IP + propofol (9.4 mg/kg) IP (Carlson et al. 2012).

Etomidate 80 µg/min IV for 3 min, then 14–26 µg/min IV.

α-chloralose 25 mg/kg IP then 75 mg/kg/h IV.

Ketamine (90)/Xylazine (9), then Ketamine (40)/Xylazine (4) every 30 min.

Pentobarbital (100, IP), then 1 mg/kg/h IV (Wang et al. 2009).

α-chloralose (120) IP + ketamine (100) IM (Rieg et al. 2004).

Thiobutabarbital (Inactin) (100) IP + ketamine (100) IP (Rieg et al. 2004).

Table 6.3: Long-Term Anesthesia in Guinea Pigs, Hamsters, and Gerbils

Guinea Pigs (mg/kg)

Propofol (26) IV then 120–150 mg/kg/h IV; consider adding opioids to decrease propofol dose.

Urethane 1.5 g/kg IV, IP.

Induction via halothane, followed by urethane 1.2 g/kg IP + α-chloralose (40, IV) (adding urethane 0.05 g/kg IV as needed). (Gardiner et al. 2007; Schwenke and Cragg 2004)

Induction via inhalant (halothane), then α-chloralose (80, IV) with vecuronium (0.5, IV) (Bootle et al. 1998).

Induction via inhalant (halothane + N$_2$O), then ketamine (5.5)/xylazine (1.4) IV, followed by ketamine 14.6 mg/kg/h/xylazine 3.7 mg/kg/h (Schwenke and Cragg 2004).

Induction via inhalant (halothane + N$_2$O), then saffan (alphaxalone-alphadolone) (4.5) then 9.75 mg/kg/h (Schwenke and Cragg 2004).

Pentobarbital (10), then 22 mg/kg/h (Schwenke and Cragg 2004).

Telazol (60, IP) + xylazine (5, IP) + butorphanol (0.1, IM) (Jacobson 2001).

Hamsters

Urethane 1–2 g/kg IP

Gerbils

Urethane 1.25 g/kg IP

α-chloralose 100 mg/kg IP

Inhalant + fentanyl 50 μg/kg IV bolus and then 50 μg/kg/h IV

6.3 neonatal anesthesia

- Neonatal rats and mice are altricial and focused mainly on feeding during the first 2 weeks.

- Both removal from the dam and siblings, and anesthesia, can cause changes that may affect research. For example, temporary separation of the neonate from its family can affect glucocorticoid levels, immune response, and neural function.

- To minimize changes resulting from separation alone, ensure the neonates continue to smell bedding and nest material from the home cage during the induction and recovery.

- Exposure of the pup to general anesthetics also results in persistent abnormal learning and fear-related behaviors and can cause neurodegeneration, even if cerebral perfusion and oxygenation and blood glucose are kept unchanged. This response is age dependent, being most severe in younger neonates.

- Neonatal pain can also lead to chronic changes in behavior, but it is unclear if altricial rodents less than 12–14 days of age are able to perceive pain. However, these young rodents do show generalized body movements and audible responses when painful stimuli are applied.

- Neonates have immature physiological functions, with little organ or energy reserves. They also have immature hepatic, renal, and brain function and so may respond to anesthetic drugs differently than older animals do.

- Inhalants are the preferred general anesthetic for neonates. Neonates may require slightly higher or lower inhalant doses than adults. Sevoflurane has been reported to affect brain maturation in 7-day-old rat pups (Makaryus et al. 2015).

- Inhalant anesthesia with isoflurane or sevoflurane can also be used in both precocial (guinea pig) and altricial neonates. Inhalant anesthesia can be used for pups of all ages, while hypothermia should not be used beyond postnatal day 7. Isoflurane has been shown to cause hypoglycemia in 7-day-old mice, according to Loepke et al., who were able to maintain normoglycemia with 0.01 ml/g 5% dextrose in saline given SC each hour. Isoflurane may also cause more neurodegeneration in 7-day-old mouse pups than sevoflurane. Neonates must maintain a high heart rate for normal cardiac output, and bradycardia in neonates may not respond to atropine because of their immature autonomic system. Hypoxia in human neonates, unlike in adults, causes initial tachycardia, followed by a normal heart rate. Hypoxia may cause apnea. Therefore, the use of 40% oxygen as a carrier gas is recommended to try to maintain normoxia. Neonates are also more prone to alveolar collapse and have a high respiratory rate along with a long expiratory time. Hypothermia and methoxyflurane were found to be effective in rat pups (Danneman and Mandrell 1997).

- Cannibalism can occur if a neonate or the dam is disturbed. Multiparous, Swiss Webster, or C3H dams have good maternal instincts and are less likely to cannibalize. Before returning the pup to the nest, remove any blood from its skin and rub the pup in nesting material or urine from the dam and littermates. Place the pup back in the middle of the litter, and monitor it closely for the next few minutes then intermittently over the next several hours. Ensure that the pup is kept in the nest and allowed to feed. Perfume, alcohol, vanilla, or another masking substance can be placed just below the dam's nostrils if necessary. Ideally, foster dams should be available in case the pup's own dam rejects it. If rejection occurs more than occasionally, consider replacing the surgicated pups with an outbred mouse's pups of the same age. After the surgery, place the surgicated pup with the foster dam for a few hours to allow it to completely recover and feed. Then replace the surgicated pup with its own dam and remove the outbred pups.

- Unless using hypothermia, during anesthesia keep the pups warm with a water circulating heating pad, incubator, or forced-air warmer. Flow-by oxygen (oxygen placed by the nose so that it flows perpendicular to the long axis of the pup) may help maintain systemic oxygenation.

- If postoperative analgesia is used, an opioid is recommended. Morphine is a respiratory depressant, so the pup should be monitored closely for signs of lethargy, lack of feeding, or cyanosis. Buprenorphine is less likely to cause respiratory suppression.

  - Morphine 5 (for rat pups) or 10 (for mouse pups) mg/kg SC or IP

  - Buprenorphine 1 mg/kg SC (for rat pups)

  - Lidocaine (2–4 mg/kg) and bupivacaine (1 mg/kg) local infiltration

- Injections should be given via a 30 gauge needle. SC injections can be given in the dorsal aspect of the neck or upper back as in adults, but the skin should be pinched closed after withdrawing the needle. For IP injections, insert SC for a short distance, then enter into the peritoneal cavity and inject. The anterior facial vein or transverse sigmoid sinus can be used for intravenous injections. Intracardiac injections may also be given in a similar manner as to adults.

6.3.1 Hypothermia

- Hypothermia can also be used to anesthetize altricial rodent neonates up to postnatal day 7, although there is controversy as to whether the resulting immobilization results in lack of pain perception. Because neural conduction stops ~ 9°C, it seems likely that deeply hypothermic pups do not experience pain at the time of deep hypothermia.

- Hypothermia can also cause persistent changes in cognitive behavior and cerebrocortical structure. The development of hypothermia may be painful, and direct contact with the pup's skin should be avoided, with a protective layer (such as a latex sleeve) always present between the pup and ice. Although hypothermia was an effective method for general anesthesia, isoflurane or sevoflurane anesthesia maintains consistent physiologic parameters and provides for faster recovery (Huss et al. 2016).

- Induction of hypothermia:

  - *Induction:* Protect the pup's skin from direct exposure to the cold with a latex sleeve/glove or plastic wrap, and place the pup in crushed ice with water (ice bath, ~35°F). Pups can be placed in glove fingers, and then into individual ice tray wells filled with ice and water. The neonate's head should not be immersed into the ice bath. Typically, induction requires 5–8 min.

  - *Maintenance:* Remove the pup from the ice bath and place in a sleeve on top of crushed ice or ice pack, and maintain the hypothermia for 15–30 min maximum. During the surgery, avoid warming the pup with incandescent or halogen lights.

  - *Recovery:* Rapid rewarming should be avoided. The pup can be placed in a nest-like blanket at 90°F–95°F environment for ~ 30 min until it begins to crawl. The author recommends providing 50%–100% O_2 during recovery. A warming light may be used. The complete recovery may take ~ 30–60 min. After recovery, the pup can be placed with the litter to minimize smells from procedure and then the whole litter can be placed with the dam.

6.4 good practice for anesthesia for small mammals

- Anesthetic depth should be evaluated via palpebral (big small mammals), pedal or ear-pinch reflexes, jaw tone, respiratory rate, heart rate, blood pressure, oxygenation, and end-tidal or arterial CO_2.

- Animals should be kept warm with a warm water circulating pad or a forced-air warming blanket. Warmed parenteral fluids and bubble wrap around extremities may help.

- Replace any blood lost with about three times the volume in SC, IV, or IP fluids. Warmed balanced fluid (lactated Ringer's solution, etc.) can be given (SC, IV, IP) at 5–10 ml/kg/h.

- Fasting is not necessary unless gastrointestinal surgery is intended. Fasting should be minimized because it may lead to ileus.

- Bland ophthalmic ointment should be placed into the animal's eyes once anesthetized and the eyelids taped closed with a gentle medical tape.

- For postoperative care, ideally analgesia should have been started before the procedure (preemptive analgesia). If this was not possible, then an analgesic should be given before the animal awakens from anesthesia. Ideally, for major surgery, a combination of opioids, local anesthetics, and nonsteroidal anti-inflammatory drugs (NSAIDs) are given initially. An NSAID or opioid alone may be adequate for 48 h after the procedure, but both can be given if necessary. Either carprofen or meloxicam are good NSAID choices. Buprenorphine is a good opioid choice for small mammals. Generally this is given every 6–12 h, but an extended release formula (Buprenorphine-SR, ZooPharm, Laramie, WY) claims efficacy for up to 72 h. No published or anecdotal data was found for guinea pigs, gerbils, or hamsters. Based on the author's experience, hamsters may be sensitive to buprenorphine; therefore, dosing should be done with care.

6.4.1 Anesthesia

- Anesthesia can be accomplished with either injectable or inhalational anesthesia.

- Injectable anesthesia can be unpredictable as its effects may vary with the individual. This unpredictability, and the need to give additional doses in some individuals, may introduce unwanted variability into the study. In addition, if a procedure is longer than anticipated or if the animal shows signs of inadequate depth of anesthesia, it is difficult to titrate additional anesthetics. The general recommendation for ketamine-based general anesthetic mixtures is to give ~1/3-1/2 the dose of ketamine only. Injectable anesthetics can be supplemented with inhalants, if necessary, for major or prolonged procedures. Even if an inhalant is not used, a mask with supplemental 100% O_2 can help maintain blood oxygenation, especially if α-2 agonists are used.

- Inhalant anesthesia (isoflurane or sevoflurane) is recommended when possible, as it allows anesthesia to be maintained for as long as necessary as well as easy and quick adjustments to be made to the anesthetic level. Ideally, a premedication containing an analgesic and sedative should be given 15–30 min before induction. This helps reduce stress on the small mammal, and it also helps reduce breath holding in response to the inhalant anesthetic. Some rodents may respond to inhalants with laryngospasm, lacrimation, salivation, and facial rubbing, which premedication can reduce. Midazolam is a good sedative and is absorbed well when given SC, IM, or IP. Acepromazine may also be used. Buprenorphine is a good choice for moderately painful procedures. Therefore, a combined injection of midazolam and buprenorphine makes a good premedication for inhalant anesthetics. Flumazenil can be used to reverse midazolam (or diazepam) if necessary, but low doses of these drugs have few side effects. Naloxone can be used to reverse opioid agonists such as morphine, hydromorphone, or fentanyl. However, very high doses of naloxone can be necessary to reverse buprenorphine, because it is a partial agonist with very high receptor affinity. Butorphanol is effective only for mild pain, but it is a more potent sedative than buprenorphine, so it can be used along with midazolam or acepromazine for more profound sedation.

- Alternatively, xylazine or the more selective α-2 agonist dexmedetomidine may also be used as premedicants. α-2 agonists may cause respiratory depression during general anesthesia, so mask supplementation with 100% O_2 can be used.

- Anticholinergics (i.e., atropine) are usually given preoperatively in some rodents (i.e., guinea pigs) because they tend to salivate profusely, especially with ketamine. Anticholinergics are thought to occasionally contribute to ileus. However, anticholinergics are generally not recommended if α-2 agonists are given. Although α-2 agonists cause bradycardia, this is a physiological response to blood pressure changes, and therefore anticholinergics are generally not recommended unless bradycardia is accompanied by hypotension. Even in that case, it may be best to reverse the α-2 agonist. Yohimbine or atipamezole can be used to reduce xylazine or dexmedetomidine if severe side effects are noted or if the animal remains anesthetized after the procedure is finished.

- When using inhalant anesthetics, equipment dead space must be minimized in these small animals. Avoid using extra tubes added onto the proximal portion of the breathing circuit. Use mainstream rather than sidestream capnographs. Use a nonrebreathing system such as Bain or Mapleson D, with a carrier gas (air or oxygen) flow of at least 150–300 ml/kg/min during spontaneous ventilation and a minimum of 500 ml/min on the flowmeter. (If intubated, a rodent $ETCO_2$ can help ensure adequate ventilation. Use a downdraft table or fume hood for mask or chamber induction to decrease excessive personnel exposure. Sedation before inhalant induction also appears to decrease stress for these generally anxious and timid animals.

6.4.2 Anesthesia in Guinea Pigs

6.4.2.1 General considerations

- Guinea pigs have strain and species differences, which can affect response to drugs.

- Laboratory guinea pigs are covered by the United States Department of Agriculture (USDA)'s Animal Welfare Act. This act contains a number of requirements, including an IACUC with specific members, appropriate anesthesia and analgesia (unless the IACUC approves otherwise), adequate veterinary care, veterinary advice about anesthesia and analgesia, a literature search for painful and distressful procedures, perioperative care, and record keeping for each individual. Expired anesthetics and analgesics must not be used. Animals must be euthanized if they experience unrelieved pain or distress not approved by the IACUC. Non-pharmaceutical-grade drugs must be specifically approved by the IACUC after adequate justification and description of how the drug will be mixed and given to preserve safety and efficacy. Instruments should be autoclaved for each individual. While a dedicated surgical room is not required, the area of the room should be dedicated to surgery, have little traffic flow, and kept clean and uncluttered. The guinea pig's surgical site should be scrubbed and draped. The surgeon should scrub his or her hands and wear a cap and mask, sterile surgical gloves, and a sterile surgical gown.

- Animals should be weighed and examined before each procedure. The large cecal volume usually present in guinea pigs may contribute to this species's reputation for difficult anesthesia, because it will influence drugs given on a mg/kg basis. Count the guinea pig's normal respiratory rate before removing it from its enclosure; then count the heart rate. Temperature may be obtained via rectal thermometer; introduce the thermometer into the perineal sac, then direct it slightly dorsally into the anus. Do not force the thermometer; if it is at the anus, it should slide into the rectum with minimal pressure. If the thermometer does not enter easily, it is not in the rectum, but rather elsewhere in the perineal sac.

- Guinea pigs often have food within their mouths, whether they have been fasted or not. This can be checked before anesthesia with a small vaginal speculum, or it can be checked and carefully cleared immediately after induction with either cotton-tipped swabs or a warm-water flush.

- The heart rate (HR) in anesthetized guinea pigs is usually low enough to be sensed by any electrocardiogram (ECG) and pulse oximeter. A pulse oximeter probe (%SpO$_2$) can be placed around a paw or distal limb; shaving may be necessary in furred areas. A Doppler flow probe may be placed over the neck in the area of carotid artery to allow the pulse rate to be heard. Respirations can be counted. Blood pressure can be obtained via the invasive method with a catheter in the carotid or femoral artery (cutdown required), or noninvasively via Doppler or oscillometric blood-pressure machine with a small cuff placed on a forelimb or hindlimb.

- Guinea pigs may become very stressed with injections, so it is best to limit the number given. Injectable drugs can be given IP in the right caudal abdomen, penetrating less than 0.5 inches into the abdomen and keeping the angle of entrance close to parallel after the body wall is entered. IM injections are usually given in the epaxial muscles or the cranial quadriceps; guinea pigs may traumatize the limbs after injection, and therefore drugs that are likely to be irritating (i.e., ketamine) should not be given in the quadriceps. The SC route (in the dorsal neck or back) can also be used and is less painful than the IM route. Use 23–25 gauge needles for injections and do not exceed 0.3 ml volume IM, 5 ml/kg bolus IV (up to 20 ml/kg if given as slow infusion), and 20–30 ml/kg volume IP or SC.

- IV access is possible via the auricular veins, cephalic vein, medial saphenous, lateral metatarsal vein, or jugular vein. A cutdown may be necessary in all but the auricular veins. If vascular access is necessary but an intravenous catheter cannot be placed, consider an intraosseous (IO) catheter. IO catheters are generally placed after anesthetic induction because placement is painful. If a critically ill guinea pig is presented, an IO catheter can be placed after injection of 1 mg/kg lidocaine locally down to the periosteum. IO catheters can be used for fluid and drug administration during the procedure, and then removed after the procedure. IO access is equivalent to intravascular and can be obtained via the proximal femur (the trochanteric fossa medial to the greater trochanter), proximal humerus (cranial greater tubercle), or proximal tibial crest. The author feels that the proximal femur is the easiest site for placement, using a 25 gauge needle with a 4-0 monofilament stainless steel suture as stylet. A small stab incision should be made in the skin after sterile preparation with povidone iodine or chlorhexidine; the needle is inserted about 1/3–1/2 the length of the femur, the stylet removed, and the catheter flushed with 0.25 ml sterile saline and an additional 1 mg/kg lidocaine. If a stylet is not used, bone marrow may plug the catheter upon placement.

- Guinea pigs are especially susceptible to antibiotic-induced enterocolitis. Because their intestinal flora is predominantly gram positive, antibiotics effective against mostly gram-positive bacteria should be avoided. However, any antibiotic has the potential to cause gastrointestinal disturbance in this species, and therefore antibiotics should not be given unless necessary. If a prophylactic surgical antibiotic is necessary, the safest choice is 5–10 mg/kg enrofloxacin given IM or (if diluted at least 50%) SC 30–60 min prior to surgery.

- Like most rodents, guinea pigs are obligate nasal breathers (Nixon 1974) but can breathe a small amount through their mouths when placed in dorsal recumbency. The mask should cover the nose if in lateral or ventral recumbency and both the nose and mouth if in dorsal recumbency. Because guinea pigs do not regurgitate, mask maintenance is generally fine. To ensure that equipment dead space is not excessive, the mask should fit over the nose (and/or mouth) only, or over the entire head (ensure the rubber flange of mask does not cause pressure on the neck).

- *Intubation:* If controlled respiration is necessary, a tracheostomy can be done if the procedure is a nonsurvival one. Controlled respiration via a laryngeal mask airway-type device has been described in other rodents but not in guinea pigs.
  - Endotracheal intubation is possible but difficult due to the guinea pig's unique pharyngeal anatomy. There are techniques described in literatures (Kramer et al. 1998; Turner et al. 1992). Guinea pigs have only a small round pharyngeal opening, called the palatal opening, formed by the soft palate's confluency with the tongue. The larynx is not seen until you have passed through the ostium. In addition, the laryngeal opening in this species is smaller than expected (Figure 6.3).
  - Endotracheal intubation is best accomplished with direct visualization of the larynx via a flexible or semirigid fiberscope or a rigid Hopkin telescope. A rodent mouth gag can be used (Figure 6.4).

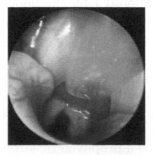

Fig. 6.3 Guinea pig laryngeal opening. (Reprinted from *Veterinary Clinics of North America: Exotic Animal Practice*, 13(2), Dan H. Johnson, DVM, endoscopic intubation of exotic companion mammals, 273–289, Copyright (2010), with permission from Elsevier.)

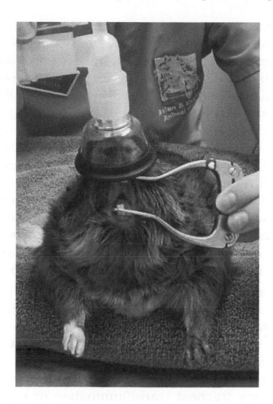

Fig. 6.4 Example of mouth gag in guinea pig. (Reprinted from *Veterinary Clinics of North America: Exotic Animal Practice*, 13(2), Dan H. Johnson, DVM, endoscopic intubation of exotic companion mammals, 273–289, Copyright (2010), with permission from Elsevier.

- Scopes with suction are ideal because they allow the evacuation of any saliva or blood that may accumulate and prevent visualization.

- For intubation, a strong suture around the upper incisors should be used. At the soft palate and pharynx, 2% lidocaine can be dropped. Cetacaine should not be used because of the potential for methemoglobinemia.

- Do not make repeated attempts at intubation, and do not intubate if signs of laryngeal trauma are seen. Depending on the animal's size, size 2–2.5 uncuffed endotracheal tubes or an 18–19 gauge IV catheter without stylet can be used.

- Ideally, clear tubes are used so that condensation can be seen through the tube. The tube's outer surface should be lubricated.

- To reduce airway resistance, the largest internal diameter possible should be used. A 2.7 mm 30° Hopkin telescope provides superior visualization but does not allow over-the-scope intubation as a 1.9 mm or 1 mm endoscope does. The Hopkin telescope is also usually too large to allow both the scope and the endotracheal tube to pass through the palatal ostium at the same time. Therefore, a 5 French rigid polypropylene catheter, plastic-coated wire, or other similar stylet may be passed initially into the larynx. The telescope is then removed, and an endotracheal tube is slid over the stylet into the larynx. If a scope is not available, intubation is possible by visualizing the larynx with an otoscope, with simultaneous transillumination of the larynx by a light source placed external to the neck. A stylet is introduced, and then the endotracheal tube slid over that as just described.

- Alternatively, place the guinea pig in dorsal recumbency (chest and head slightly elevated) and tape the maxilla to the table. Place laryngoscope into the diastema, parallel to the mouth. Introduce a stylet and place it blindly into the trachea during inspiration; then insert an endotracheal tube over this. Proper positioning should be confirmed via tracheal transillumination or capnography if possible. Small-diameter endotracheal tubes are prone to blockage with secretions; therefore, monitor for this closely. Atropine can reduce salivary secretions but may also make secretions more viscous and therefore make tracheal plugs more difficult to clear.

- After surgery, guinea pigs should be monitored for signs of pain, which include piloerection or unkempt fur, a hunched appearance, anorexia, aggression, vocalization when the surgical site is palpated, or facing into a corner or the back of the cage. Unfortunately, because guinea pigs are stoic, pain signs may not be seen. Opioids themselves may cause ileus and lack of appetite, but this is uncommon with buprenorphine in guinea pigs. It is imperative in this species to control pain, which may cause ileus. Grass hay should be offered to all guinea pigs postoperatively. Irradiated timothy hay (BioServ, Frenchtown, NJ) is available, or hay may be autoclaved and then dried before presentation to the guinea pig. Animals should be weighed prior to the procedure and then at least once daily after the procedure until weight stabilizes. If anorexia and weight loss continue despite adequate analgesia, special foods may be offered to tempt appetite. The author has had success with alfalfa-based Critical Care (Oxbow Animal Health, Murdock, NE), mixed with about 25% plain rice baby food, which provides increased simple carbohydrates and makes the Critical Care easy to syringe feed. Alternatively, if increased calories are needed, the Critical Care (Figure 6.5) powder may be mixed with any liquid diet manufactured for dogs or humans. Guinea pigs tolerate syringe feeding well, and feeding should be started as soon as anorexia is noted. Critical Care can also be mixed with water until it can be formed into several small balls, which can be placed into the guinea pig's cage. Washed leafy vegetables and small amounts of fruit can be offered in a nonbarrier facility. Extra vitamin C (30–100 mg/d) should be offered to all anorexic guinea pigs. Although guinea pigs often initially decline new foods, they usually soon begin to like the taste of oral vitamin C solution. Therefore, it can be given as a "treat" before syringe feeding, or it may be mixed into the food.

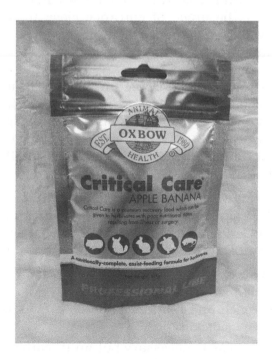

Fig. 6.5 Sample of critical care food (Murdock, NE).

6.4.3 Anesthesia in Hamsters

- Hamsters have strain, species, and genus differences, which can affect response to drugs.

- Unlike most mice and rats, laboratory hamsters are covered by the USDA's Animal Welfare Act. (See details for guinea pigs.)

- Hamsters have a reputation for being feisty and for frequently biting. It is worthwhile to handle the hamster several times before the procedure, giving positive attention and small treats, before handling for the potentially distressful procedure. Unlike many rodents, hamsters are generally solitary animals, and they are best housed alone. If they are sleeping, make sure to wake them up before handling them. To pick up, scruff by the large amount of loose skin they have at the back and neck. Hamsters can be restrained by hand, or by rigid or "snuggle" (canvas with Velcro fasteners) restrainers.

- Animals should be weighed and examined before each procedure.

- Injectable drugs can be given IP in the right caudal abdomen, penetrating less than 0.5 inches into the abdomen and keeping the angle of entrance close to parallel after the body wall is entered. Male hamsters have large testes, so use caution not to accidentally penetrate the testes during IP injections. IM injections are usually given in the cranial quadriceps. The SC route (in the dorsal neck or back) can also be used and is less painful than the IM route, but absorption and time to onset will be longer and peak efficacy may be less. Use 23–25 gauge needles for injections, and do not exceed 0.1 ml volume IM, 0.3 ml volume IV, and 3 ml volume IP or SC.

- Hamsters can be susceptible to antibiotic-induced enterocolitis, and tetracyclines have been reported to cause deaths. If a prophylactic surgical antibiotic is necessary, the safest choice is 5–10 mg/kg enrofloxacin given IM or (if diluted at least 50%) SC 30–60 min prior to surgery.

- Bland ophthalmic ointment should be placed into the hamster's eyes once anesthetized and the eyelids taped closed with a gentle medical tape.

- If vascular access is necessary but an IV catheter cannot be placed, consider an IO catheter.

- Like most rodents, hamsters breathe mostly through their nose and are unable to breathe much through their mouths except when placed in dorsal recumbency.

6.4.4 Anesthesia in Gerbils

- Gerbils have strain and species differences, which can affect response to drugs.

- Unlike most mice and rats, laboratory gerbils are covered by the USDA's Animal Welfare Act (see guinea pigs).

- Gerbils are generally docile, but like most animals, they will bite if sufficiently annoyed or frightened. They can be picked up or momentarily restrained by the base of the tail, then gently cupped in the other hand with the handler's thumb under the mandible or under one forelimb and the handler's palm supporting the animal's rump. Alternatively, they can be scruffed by the skin at the back of their neck with the handler's thumb and forefinger, with the handler's palm supporting the animal's rump. Gerbils can be restrained by hand, or by rigid or "snuggle" (canvas with Velcro fasteners) restrainers. Gerbils prefer to be held vertically rather than in dorsal recumbency in cases where the ventral portion of their body must be manipulated. The handler should be aware that some gerbils seizure, especially if handled extensively or in a noisy environment. If this occurs, place the gerbil back in its home cage in a quiet environment and allow it to recover. Seizures are more common in some strains and usually do not occur before 6 weeks of age. Typically, seizures last less than 90 s and anticonvulsant medication is not needed. Frequent handling of gerbils, starting from a young age, may decrease seizure tendency.

- Gerbils can be susceptible to antibiotic-induced enterocolitis. Dihydrostreptomycin and procaine are contraindicated (Mitchell and Tully Jr. 2008). If a prophylactic surgical antibiotic is necessary, the safest choice is 5–10 mg/kg enrofloxacin given IM or (if diluted at least 50%) SC 30–60 min prior to surgery.

- Gerbils breathe mostly through their nose and are unable to breathe much through their mouths except when placed in dorsal recumbency.

Commonly used anesthetic drugs and combinations are shown in Table 6.4.

Table 6.4: Drug Dosages

Drug (mg/kg)	Rat	Mouse	Guinea Pig	Hamster	Gerbil
Acepromazine	2.5SC, IM, IP	2-5SC, IP	0.5-1SC, IP	3-5SC, IP	3-5SC, IP
Diazepam	2.5-5IM, IP	3-5IM, IP	2.5IM, IP	3-5IM, IP	3-5IM, IP
Midazolam	1-2IM, IP	1-2IM, IP	1-2IM, IP	1-2IM, IP	1-2IM, IP
Dexmedetomidine	0.015-0.05SC	0.015-0.055SC	0.25IM, IP		
Xylazine	5SC, IM	5SC, IM		5SC, IM	
Ketamine	50-100IM, IP	100-200IM	40-100IM, IP	0.04SC, IM	0.04SC, IM
Atropine	0.04SC, IM	0.04SC, IM	0.04SC, IM	0.04SC, IM	0.04SC, IM
Alfaxalone			40IM, IP		
Ketamine/dexmedetomidine	60-75/0.12-0.25IP	100/0.25IP	40/0.25IP		
Ketamine/xylazine	40-100/5-10 IP	80-100/10 IP	40/5 IP	50-200/5-10 IP	50/2 IP
Ketamine/acepromazine	50-150/2.5-5 IP	50-150/2.5-5 IP	100/5 IP	50-150/2.5-5 IP	
Ketamine/xylazine/acepromazine	100/2.5/2.5IP	100/2.5/2.5IP			
Telazol	50-80IP, IV	50-80IP	20-40IM, IP	50-80IP	50-80IP
Telazol/xylazine/butorphanol			60/5/0.1 IP		
Ketamine/midazolam	5-10/0.25-0.5 IP		5-15/0.5-1 IP		
Ketamine/diazepam	40-100/3-5IP, IV	40-150/3-5 IP		50-150/5 IP	40-150/3-5 IP
Propofol	2-5IV	7.5-26IV, IP			7.5-26IV, IP
Pentobarbital	30-60IP	50-90IP	15-40IP		60IP
Urethane	100IP		1500IP		
Pancuronium	2IV		0.06IV		
Vecuronium	1IV				
Yohimbine (reverse xylazine)	0.1-0.25SC, IM				
Atipamezole (reverse dexmedetomidine)	0.02-0.06IM, IV				

Source: Dosages are in mg/kg unless otherwise indicated. (Jacobson, C. *Lab. Anim.*, 35(3), 271–276, 2001; Wenger, S., *J. Exot. Pet Med.*, 21(1), 7–16, 2012; Brown, C. and T. M. Donnelly, *Ferrets, Rabbits and Rodents, Clinical Medicine and Surgery*, Saunders, 2012; Miller, A. L. and C. A. Richardson, *Vet. Clin. North. Am. Exot. Anim. Pract.*, 14(1), 81–92, 2011; Flecknell, P. A. and A. A. Thomas, *Veterinary Anesthesia and Analgesia, the Fifth Edition of Lumb and Jones*, Wiley, Ames, IA, 2015; Steffey, E. P. and K. R. Mama, *Veterinary Anesthesia, the Fifth Edition of Lumb and Jones*, Wiley, Ames, IA, 2007; Flecknell, P.A. et al., *Essentials of Small Animal Anesthesia & Analgesia*, Lippincott Williams & Wilkins, London, 1999; Richardson, C. and P. Flecknell, *BSAVA Manual of Rodents and Ferrets*, BSAVA, Gloucester, UK, 2009) some are from the author's experiences.

references

Alstrup, A. K. and D. F. Smith. 2013. Anaesthesia for positron emission tomography scanning of animal brains. *Lab. Anim.* 47(1):12–18.

Bootle, D. J., J. J. Adcock, and A. G. Ramage. 1998. The role of central 5-HT receptors in the bronchoconstriction evoked by inhaled capsaicin in anaesthetised guinea-pigs. *Neuropharmacology* 37(2):243–250.

Brown, C. and T. M. Donnelly. 2012. Disease problems of small rodents. In *Ferrets, Rabbits and Rodents, Clinical Medicine and Surgery*, eds. K. E. Quesenberry and J. W. Carpenter, 356–369: Saunders.

Cabanero, D., A. Campillo, E. Celerier, A. Romero, and M. M. Puig. 2009. Pronociceptive effects of remifentanil in a mouse model of postsurgical pain: Effect of a second surgery. *Anesthesiology* 111(6):1334–1345.

Carlson, A. P., R. E. Carter, and C. W. Shuttleworth. 2012. Vascular, electrophysiological, and metabolic consequences of cortical spreading depression in a mouse model of simulated neurosurgical conditions. *Neurol. Res.* 34(3):223–231.

Danneman, P. J. and T. D. Mandrell. 1997. Evaluation of five agents/methods for anesthesia of neonatal rats. *Lab. Anim. Sci.* 47(4):386–395.

Driehuys, B., J. Nouls, A. Badea, et al. 2008. Small animal imaging with magnetic resonance microscopy. *ILAR J.* 49(1):35–53.

Ferrari, L., G. Turrini, V. Crestan, et al. 2012. A robust experimental protocol for pharmacological fMRI in rats and mice. *J. Neurosci. Methods* 204(1):9–18.

Flecknell, P. A., C. A. Richardson, and A. Popovic. 1999. Anesthesia and immobilization of small animals. In *Essentials of Small Animal Anesthesia & Analgesia*, eds. K. A. Grimm, W. J. Tranquilli, and L. A. Lamont. London: Lippincott Williams & Wilkins.

Flecknell, P. A. and A. A. Thomas. 2015. Comparative anesthesia and analgesia of laboratory animals. In *Veterinary Anesthesia and Analgesia, the Fifth Edition of Lumb and Jones*, 5th edition, 754–763. eds. K. A. Grimm, L. A. Lamont, W. J. Tranquilli, S. A. Greens and S. A. Robertson. Ames, IA: Wiley.

Gardiner, J. C., G. McMurray, and S. Westbrook. 2007. A bladder-cooling reflex in the anaesthetised guinea-pig: A model of the positive clinical ice-water test. *J. Pharmacol. Toxicol. Methods* 55(2):184–192.

Hanusch, C., S. Hoeger, and G. C. Beck. 2007. Anaesthesia of small rodents during magnetic resonance imaging. *Methods* 43(1):68–78.

Hughes, E. W., R. L. Martin-Body, I. H. Sarelius, and J. D. Sinclair. 1982. Effects of urethane-chloralose anaesthesia on respiration in the rat. *Clin. Exp. Pharmacol. Physiol.* 9(2):119–127.

Huss, M. K., H. H. Chum, A. G. Chang, K. Jampachairsi, and C. Pacharinsak. 2016. The physiologic effects of isoflurane, sevoflurane, and hypothermia used for anesthesia in neonatal rats (*Rattus norvegicus*). *J. Am. Assoc. Lab. Anim. Sci.* 55(1):83–88.

Ismaiel, N. M., R. Chankalal, J. Zhou, and D. Henzler. 2012. Using remifentanil in mechanically ventilated rats to provide continuous analgosedation. *J. Am. Assoc. Lab. Anim. Sci.* 51(1):58–62.

Jacobson, C. 2001. A novel anaesthetic regimen for surgical procedures in guineapigs. *Lab. Anim.* 35(3):271–276.

Kramer, K., J. A. Grimbergen, D. J. van Iperen, K. van Altena, H. P. Voss, and A. Bast. 1998. Oral endotrachael intubation of guineapigs. *Lab. Anim.* 32(2):162–164.

Makaryus, R., H. Lee, T. Feng, et al. 2015. Brain maturation in neonatal rodents is impeded by sevoflurane anesthesia. *Anesthesiology* 123(3):557–568.

McConville, P. 2011. Small animal preparation and handling in MRI. *Methods Mol. Biol.* 771:89–113.

Miller, A. L. and C. A. Richardson. 2011. Rodent analgesia. *Vet. Clin. North. Am. Exot. Anim. Pract.* 14(1):81–92.

Mitchell, M. and T. N. Tully Jr. 2008. Hamsters and gerbils. In *Manual of Exotic Pet Practice*, 406–432. St. Louis, MO: Saunders Elsevier.

Nixon, J. M. 1974. Breathing pattern in the guinea-pig. *Lab. Anim.* 8(1):71–77.

Rajda, C., D. Dereczyk, and P. Kunkel. 2008. Propofol infusion syndrome. *J. Trauma Nurs.* 15(3):118–122.

Richardson, C. and P. Flecknell. 2009. Rodents: Anaesthesia and analgesia. In *BSAVA Manual of Rodents and Ferrets*, eds. E. Keeble and A. Meredith, 63–72. Gloucester, UK: BSAVA.

Rieg, T., K. Richter, H. Osswald, and V. Vallon. 2004. Kidney function in mice: Thiobutabarbital versus alpha-chloralose anesthesia. *Naunyn Schmiedebergs Arch. Pharmacol.* 370(4):320–323.

Schwenke, D. O. and P. A. Cragg. 2004. Comparison of the depressive effects of four anesthetic regimens on ventilatory and cardiovascular variables in the guinea pig. *Comp. Med.* 54(1):77–85.

Simpson, D. P. 1997. Prolonged (12 hours) intravenous anesthesia in the rat. *Lab. Anim. Sci.* 47(5):519–523.

Steffey, E. P. and K. R. Mama. 2007. Inhalation anesthetics. In *Veterinary Anesthesia, the Fifth Edition of Lumb and Jones*, 3, 297–329, eds. K. A. Grimm, L. A. Lamont, W. J. Tranquilli, S. A. Greens, and S. A. Robertson 3. Ames, IA: Wiley.

Tung, A., S. Herrera, C. A. Fornal, and B. L. Jacobs. 2008. The effect of prolonged anesthesia with isoflurane, propofol, dexmedetomidine, or ketamine on neural cell proliferation in the adult rat. *Anesth. Analg.* 106(6):1772–1777.

Tung, A., J. P. Lynch, and W. B. Mendelson. 2001. Prolonged sedation with propofol in the rat does not result in sleep deprivation. *Anesth. Analg.* 92(5):1232–1236.

Turner, M. A., P. Thomas, and D. J. Sheridan. 1992. An improved method for direct laryngeal intubation in the guineapig. *Lab. Anim.* 26(1):25–28.

van den Brink, I., F. van de Pol, M. Vancker, et al. 2013. Mechanical ventilation of mice under general anesthesia during experimental procedures. *Lab. Anim. (NY)* 42(7):253–257.

Wang, D. S., M. D. Dake, J. M. Park, and M. D. Kuo. 2009. Molecular imaging: A primer for interventionalists and imagers. *J. Vasc. Interv. Radiol.* 20(7 Suppl):S505–S522.

Wenger, S. 2012. Anesthesia and analgesia in rabbits and rodents. *J. Exot. Pet Med.* 21(1):7–16.

pain management

Patricia Foley

contents

7.1 overview

- Controlling and preventing pain in research animals is both an ethical and a scientific imperative.

- Guiding principles for use of laboratory animals throughout the world share this as one of their common tenets.
- In the United States, Institutional Animal Care and Use Committees (IACUCs) are required to consider procedures that may cause pain or distress during its review of animal use protocols and to ensure that anesthetics and analgesics are used to prevent or minimize pain and distress unless scientifically justified otherwise (Animal Welfare Act 1966; Interagency Research Animal Committee 1985; National Research Council 2011; Public Health Service [PHS] 1996).
- It has been shown in human patients that effective postoperative pain management decreases morbidity (Ballantyne et al. 1998).
- The American College of Laboratory Animal Medicine published a position paper stating that "Procedures expected to cause more than slight or momentary pain require the appropriate use of pain-relieving measures unless scientifically justified in an approved animal care and use protocol" (Kohn et al. 2007).
- Formulation of pain management plans must include as key considerations:
  - The needs of the research study
  - Potential impact of any pharmacological intervention on the scientific outcomes
  - Preventive and multimodal analgesia

7.2 physiology of pain

- The International Association for the Study of Pain defines pain as "An unpleasant sensory and emotional experience associated with actual or potential damage or described in terms of such damage" (Merskey and Bogduk 1994).
- Pain can be classified as acute or chronic, and either visceral or somatic.
- It can further be categorized as physiologic, pathologic, or neurogenic.
- Perception of pain requires involvement of four components (Perkowski and Wetmore 2006) (Figure 7.1).

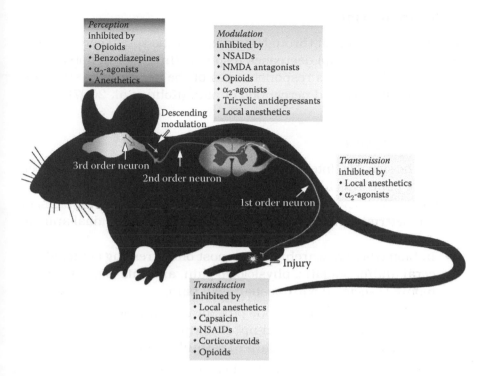

Perception
inhibited by
• Opioids
• Benzodiazepines
• α_2-agonists
• Anesthetics

Modulation
inhibited by
• NSAIDs
• NMDA antagonists
• Opioids
• α_2-agonists
• Tricyclic antidepressants
• Local anesthetics

Descending
modulation

3rd order neuron

2nd order neuron

1st order neuron

Transmission
inhibited by
• Local anesthetics
• α_2-agonists

Injury

Transduction
inhibited by
• Local anesthetics
• Capsaicin
• NSAIDs
• Corticosteroids
• Opioids

Fig. 7.1 A graphic representation of pain pathways and the four components of pain perception: transduction, transmission, modulation, and perception. Different sites of action of analgesic drugs are also shown. (Courtesy of Cholawat Pacharinsak.)

7.2.1 Transduction

A noxious stimulus is converted into an electrical signal by activation of nociceptors. A peripheral noxious stimulus might be thermal, mechanical, or chemical (distention/ischemia for visceral nociceptors).

7.2.2 Transmission

The electrical signals relay from the nociceptors to the spinal cord. Transmission continues from the spinal cord along ascending pathways to the thalamus, reticular formation, brainstem, and somatosensory cortex.

7.2.3 Modulation

The message is relayed through synapses in the spinal cord. Central sensitization (wind-up) decreases response threshold in dorsal horn neurons and increases responsiveness of the neuron to a stimulus, leading to an increased perception of pain (Kohn et al. 2007).

7.2.4 Perception

Finally, the signal is "integrated" in the cortex into an individualized perception of pain. This final process is the hardest to evaluate objectively in animals. We know they feel pain, but we cannot yet, in any systematic manner, determine what that truly looks like for the individual animal.

- In biomedical research, we are most often treating acute, visceral and/or somatic, physiologic pain, and this is most commonly associated with postsurgical pain.

- Pathologic pain may also be present in a number of research models as a result of "wind-up" resulting in increased sensitivity to nonpainful (allodynia) and painful stimuli (hyperalgesia).

- Some cancer models and other inflammatory models, for example, arthritis, may result in chronic pain that should be addressed by some form of clinical intervention.

- In addition to the potential for animal suffering, pain is a stressor that initiates a host of negative clinical effects, including catecholamine release,; stress hormone release; inappetance; immobility, which may slow down bone healing; and insomnia (Grandin and Deesing 2002; McGuire et al. 2006; Muir 2002). Pain can contribute to immunosuppression (Sternberg and Liebeskind 1995).

- Measurable responses to painful stimuli differ by species, strain, and sex (Wright-Williams et al. 2007; Wilson and Mogil 2001; Mogil and Bailey 2010).

7.3 pharmacology

- Early recognition of signs of pain will help better manage clinical status.

- Preventive analgesia is considered to be good practice (Dahl and Kehlet 2011) by providing analgesia before (preemptive), during, and after surgery. This will avoid or minimize the chance of the patient experiencing allodynia and/or hyperalgesia.

- Multimodal analgesia, using drugs of different classes or techniques, should be considered, particularly for surgical procedures that are expected to induce moderate to severe levels of pain. Different drugs act on different parts of the pain pathways and work in a synergistic fashion to control pain while minimizing adverse drug effects.

- All of these drugs demonstrate species, strain, and sex differences in potency and duration of effect. This has been noted in a number of studies (Mogil and Bailey 2010; Fagioli et al. 1990; Wilson and Mogil 2001).

7.3.1 Commonly Used Analgesics for Rodents

7.3.1.1 Opioids

- There are three primary opioid receptors: μ, δ, and κ receptors. Opioid agonists exert their effects at one or more of these receptor sites.

- Opioids differ widely in their potency, duration of action, and adverse effects.
    - Side effects of opioids (Perkowski and Wetmore 2006) given in dose-dependent manner
        - Sedation
        - Respiratory depression (seen primarily with mu-agonists)
        - Gastrointestinal hypomotility and possible ileus
        - Inappetance, nausea
        - Pica (most likely due to nausea; uncommon but reported with buprenorphine use; see later)
        - Excitement or dysphoria
        - Hypotension (secondary to histamine release)
        - Development of tolerance

7.3.1.1.1 Full μ agonists

- Morphine:
    - Its short duration of action makes it of limited utility for postsurgical pain prevention/relief.

- Its duration is 2–3 h IM/SC.
- It interacts with a number of other anesthetics and analgesics and may reduce the necessary dose of both morphine and the interacting drug.
- Respiratory depression may be seen as well as gastrointestinal effects.
- It causes histamine release in many species.
- Liposomal preparations have been used to extend duration of action. Commercially available formulations include DepoDur® and MST Continus®, which was tested for efficacy in rats by Leach et al. (2010). Experimental formulations have been tested in rats and have shown the ability to provide analgesia for up to 7 days (Smith et al. 2003).
- Development of tolerance is possible with repeated dosing, and thus morphine is not a good candidate for provision of chronic pain relief (Tokuyama et al. 1998).
- Fentanyl
  - Very short duration of action (5–15 min when given IV).
  - Potency: ~100X more potent than morphine.
  - No ceiling effect.
  - Analgesic effect dose dependent.
  - Useful intraoperatively but has little clinical use as postoperative analgesic at this time.
  - Profound respiratory depression may be seen in dose-dependent manner (Harper et al. 1976).
- Hydromorphone
  - Its duration of action is similar to morphine.
  - It is ~ 8× more potent than morphine.
  - Liposomal formulations have been tested in rats (Schmidt et al. 2011).

7.3.1.1.2 Partial μ agonists

- Buprenorphine
  - Most commonly used opioid analgesic in laboratory rodents.
  - Partial agonist at both μ and κ receptors (Lutfy and Cowan 2004).

- While prevailing recommendations are to dose buprenorphine at 8–12 h intervals, there is evidence suggesting duration of action is actually shorter. One study (Gades et al. 2000) achieved 6–8 h analgesia in rats (0.5 mg/kg SC) and only 3–5 h in mice (2.0 mg/kg SC).

- As a partial agonist, exerts a ceiling effect (bell shaped dose response curve).

- Analgesic potency may not be sufficient for severe pain.

- Pica (eating bedding material) reported occasionally in Sprague Dawley rats (Clark et al. 1997; Jacobson 2000). If this occurs, suggest switching to a nonopioid analgesic for those experimental cohorts.

- Oral dosing has been reported to be effective at high doses of 0.4 mg/kg (Roughan and Flecknell 2004; Flecknell 1996; Goldkuhl et al. 2010). However, absorption rate will be variable and also affected by surgical intervention and anesthesia.

- A sustained release formulation (SR-Buprenorphine) is now commercially available in the United States. Several studies have explored its potential as a postoperative analgesic in rats (Foley et al. 2011; Chum et al. 2014) and mice (Carbone et al. 2012; Jirkof et al. 2015; Kendall et al. 2016; Clark 2014; Healy et al. 2014).

7.3.1.1.3 κ *Agonist:* μ *antagonist*

- Butorphanol
  - Duration of action: Short~ 1–2 h
  - Mild sedative effect (Gades et al. 2000)
  - Effective in treating (by nasal inhalation route) sciatic nerve injury induced neuropathic pain in rats (Chen et al. 2014) and tail flick test in mice (Izer et al. 2014)

7.3.1.2 Nonsteroidal Anti-Inflammatory Drugs (NSAIDs)

- They have analgesic, antipyretic, and anti-inflammatory activities.

- They exert their analgesic action primarily by modulating the peripheral and central responses to noxious stimuli.

- Prostaglandins (PGs) act in several ways to increase pain (Liles and Flecknell 1992):

- Bind to sensory nerve endings and increase discharge of impulses
- Sensitize peripheral nociceptors
- Proinflammatory actions such as vasodilation, increasing vascular permeability, and chemotaxis of neutrophils
- NSAIDS work as cyclooxygenase (COX) inhibitors to decrease inflammation and block production of PGs as well as kinins (Kohn et al. 2007).
- NSAIDs with predominantly cyclooxygenase 2 (COX-2) activity are preferable. COX-2 activity increases in response to inflammation and is found primarily in inflammatory cells, peripheral nerves, and the central nervous system (CNS).
- COX-1 is constitutively expressed and is present in a broad array of tissues, including the stomach mucosa, liver, kidneys, and platelets. Most of the adverse clinical effects associated with COX inhibitors are those that inhibit COX-1 activity.
- Considered to be useful for mild to moderate pain procedures as a general rule, but more recent studies (Flecknell et al. 1999; Roughan and Flecknell 2001, 2004) show them to be equally effective as opioid analgesics in controlling postsurgical pain for a variety of procedures.
- Anti-inflammatory effects are responsible for their ability to prevent and relieve pain resulting from surgical trauma to tissues (Liles and Flecknell 1992).
- Antipyretic action operates by inhibition of PGE2 production in the hypothalamus (Saper and Breder 1994; Dascombe 1985).
- Side effects
  - Adverse effects include
    - Gastrointestinal ulceration and hemorrhage
    - Kidney toxicity
    - Liver toxicity
    - Decreased platelet function
  - Side effects are rarely seen in laboratory rodents because therapy rarely lasts more than 2–3 days.
  - Most commonly used NSAIDs in rats and mice are
    - Carprofen 2–5 mg/kg
    - Ketoprofen 2–5 mg/kg
    - Meloxicam 1–2 mg/kg

7.3.1.3 Local anesthetics

- Use of a local anesthetic for incisional, topical, or regional blockade can be a useful adjunct to systemic analgesics (Kohn et al. 2007).

- Mechanism of action is through reversible binding to sodium channels in peripheral nerves, blocking sodium influx and thus preventing generation of action potentials (Flecknell 1996).

- The two general classes of drugs are amino esters or amino amides.

  - Amides: Lidocaine, mepivacaine, bupivacaine, and ropivacaine.

  - Esters: Cocaine, procaine, tetracaine, and benzocaine. These are more likely to be used for topical applications such as a benzocaine gel or spray. *Caution to ensure that the animal does not ingest the drug is necessary as toxicity can occur and result in both cardiac and neurologic effects.*

- Epinephrine is often added to the solutions to counteract the vasodilatory property of local anesthetics. Epinephrine helps slow absorption rate and extend duration of action. Epinephrine may be absorbed into systemic circulation causing side effects.

- In rodents, local anesthetics are most commonly used subcutaneously (SC) to infiltrate incision areas and provide local sensory blockade.

  - Lidocaine and bupivacaine or good choices for incisional blockade depending on duration of action and potency desired.

  - Lidocaine administered as a 1% or 2% solution (should be further diluted for use in mice) provides a shorter duration of action (1.5–3 h in dogs and cats) (Perkowski and Wetmore 2006) but has a faster onset and is therefore useful for intraoperative blockade.

  - Bupivacaine is reported to provide 4–8 h of neural blockade in companion animals and is therefore a good choice for providing postoperative analgesia at the surgical site (Shilo et al. 2010). The duration of action is likely to be less in rodents due to their smaller size and higher metabolic rate. It is recommended that it be administered at

0.25% for larger rodents (rats, hamsters, guinea pigs) and diluted to 0.025%–0.1% for mice to avoid reaching toxic concentrations.

- Ropivacaine HCl is expected to have similar or longer duration of action to bupivacaine and provided close to 6 h sensory and motor blockade in a rat sciatic nerve blockade model (Foley et al. 2013).

7.3.1.4 Other drugs and analgesic adjuvants

7.3.1.4.1 Tramadol

- It works as both a weak μ-opioid agonist and as a serotonin and noradrenaline uptake inhibitor.
- It is formulated for both parenteral and oral dosing.
- It is gaining favor in clinical practices for treating pain in dogs and cats.
- Several publications address its use in rats and mice:
    - Affaitati et al. found it effective in a visceral pain model (Affaitati et al. 2002).
    - A dose of 12.5 mg/kg intraperitoneally (IP) in rats appeared effective postlaparotomy (Zegre Cannon et al. 2011).
    - Mice treated with Tramadol 20 mg/kg SC after embryo transfer (ET) surgery had no adverse impact on ET success rate (Koutroli et al. 2014)
- Tramadol 10 mg/kg IP was ineffective on its own for analgesia in an incisional pain model in rats (McKeon et al. 2011).

7.3.1.4.2 Gabapentin

- Antiepileptic drug.
- Structural analog of γ-aminobutyric acid (GABA).
- Analgesic properties are poorly understood.
- Often considered useful for chronic and/or neuropathic pain (Mao and Chen 2000).
- In combination with tramadol, decreased thermal hyperalgesia in a rat incisional model of pain (McKeon et al. 2011).

7.3.1.4.3 α-2 agonists

- They provide analgesia by acting on receptors in the spinal cord and brain stem.

- Sedation and cardiovascular effects limit their utility in pain management.
- Some studies show that very low doses may provide analgesia while minimizing the negative cardiovascular effects (Pascoe 2005).

7.3.1.4.4 Ketamine

- NMDA receptor antagonist.
- Potentially useful adjunct analgesic in combination with an opioid, for example, morphine (Redwine and Trujillo 2003).
- Should be administered at subanesthetic dose.
- Prevents "wind-up."

7.3.1.4.5 Acetaminophen

- Works both centrally and peripherally through mechanisms other than COX inhibition.
- May be a good candidate for models of inflammation where COX1 and 2 inhibitors would be contraindicated (Abbott and Hellemans 2000).

Table 7.1 presents suggested dosages.

Table 7.1: SUGGESTED DRUG DOSAGES FOR ANALGESIA IN LABORATORY MOUSE AND RAT

Drug	Mouse (mg/kg dose)	Rat (mg/kg dose)
Buprenorphine	0.05–0.1 SC q6–12 h	0.01–0.05 SC q 8–12 h
Buprenorphine SR	0.6–1.2 SC q 24–72 h	0.3–1.2 SC q 48–72 h
Morphine	2–10 SC q 2–3 h	2–10 SC q 2–3 h
Oxymorphone	0.2–0.5 SC q 4 h	0.2–0.5 SC q 4 h
Pethidine	10–20 SC q 2–3 h	10–20 SC q 2–3 h
Butorphanol	1–5 SC, IP q 2–4 h	1–2 SC, IP q 2–4 h
Carprofen	2–5 SC q 12–24 h (10 mg/kg reported use)	2–5 SC q 12–24 h (10 mg/kg reported use)
Ketoprofen	2–5 SC q 24 h	2–5 SC q 24 h
Meloxicam	1–2 SC q 12 h	1–2 SC q 12 h
Tramadol	5–20 q 6–12 h	5–10 q 6–12 h
Acetaminophen	100–200 PO	50–100 PO
1–2% Lidocaine	2–4 Local infiltration	2–4 Local infiltration
0.5% Bupivacaine	1–2 Local infiltration	1–2 Local infiltration
0.2% Ropivacaine	1–2 Local infiltration	1–2 Local infiltration

7.4 pain assessment

7.4.1 Overview

- Pain management is only truly effective when assessment of the individual animal for evidence of pain is included in the management plan.

- Pain assessment in rodents is complicated by the fact that because they are a prey species, signs of pain or distress may be concealed.

- General assessments of animal behavior may detect indications of suboptimal well-being or health but are not necessarily specific to pain.

- Recent work is revealing both pain-specific behaviors (Roughan and Flecknell 2000, 2004; Leach et al. 2012) and facial expressions that correlate with pain experience (Langford et al. 2010; Matsumiya et al. 2012; Sotocinal et al. 2011).

- Behaviors associated with pain have been characterized in laparotomy studies and can also be seen after IP injection of acetic acid as shown in Figure 7.2.

- Analysis of mouse or rat facial grimace scale requires high-quality images (for example see Figure 7.3), is difficult to perform in nonalbino animals, and has not yet been experimentally verified as a reliable cage side assessment tool; however, incorporation of various aspects of the grimace scale is gaining traction and being incorporated into pain-assessment plans.

- Table 7.2 characterizes different clinical signs that can be used to monitor for pain in rodents.

- Effective pain assessment requires cage side observation with minimal disruption to the animal for 5–10 min.

- Pain assessment can be useful to determine efficacy of treatment and to optimize titration of dose and duration of administration.

7.4.2 Use of Pain Scales

- Several groups have tried developing pain scales that would be useful for assessing pain in rodents in a more objective manner.

- Pain scales must be validated against varying "known" measures of pain and must be easily taught to confer consistency between different raters.

Fig. 7.2 A mouse showing abdominal pressing after an IP injection of acetic acid. This behavior is indicative of abdominal pain in mice and rats. (Courtesy of Patricia Foley.)

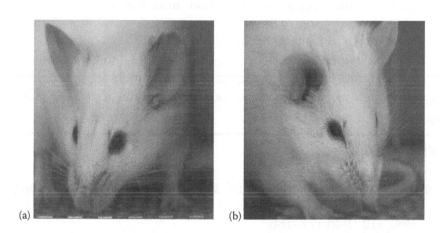

(a)   (b)

Fig. 7.3 Mouse facial images captured from video footage to assess facial grimace scale. This is a useful tool both for research purposes when evaluating analgesic efficacy and/or pain studies but also for clinical assessment of pain. (Courtesy of Patricia Foley.)

- Published studies utilizing a pain scale for assessment in mice or rats used parameters such as posture, measures of physical condition, food and water consumption, grooming, activity level/locomotion, and other behaviors (Roughan and

Table 7.2: CLINICAL OBSERVATIONS ASSOCIATED WITH PAIN IN RODENTS

General Observations	Behavioral Signs	Grimace Scale
Decreased food and water intake	Aggression/biting	Orbital tightening
Weight loss	Twitching, trembling	Nose bulge
Changes in respiratory rate and/or character	Decreased grooming activity	Cheek bulge
Increased or decreased activity	Abdominal press	Change in whisker position
Abnormal posture	Arching of back	Change in ear carriage and position
Piloerection	Writhing	
Unkempt appearance	Staggering	
Inflammation at surgical site		
Porphyrin staining (primarily rats)		
Dehydration/sunken eyes		

Flecknell 2000; Gillingham et al. 2001; van Loo et al. 1997; Liles et al. 1998).

- Most studies reported that food and water intake, body weight, and some behaviors were useful correlates of post-surgical pain (Stasiak et al. 2003).

  - These parameters are not specific to pain alone and can also be influenced by analgesic administration (e.g., sedation, appetite suppression), so they must be interpreted carefully.

references

Abbott, F. V. and K. G. Hellemans. 2000. Phenacetin, acetaminophen and dipyrone: Analgesic and rewarding effects. *Behav. Brain Res.* 112 (1–2):177–186.

Affaitati, G., M. A. Giamberardino, R. Lerza, D. Lapenna, S. De Laurentis, and L. Vecchiet. 2002. Effects of tramadol on behavioural indicators of colic pain in a rat model of ureteral calculosis. *Fundam. Clin. Pharmacol.* 16 (1):23–30.

Animal Welfare Act:. Public Law 89-544, August 24, 1966, as Amended (P.L. 91-579, P.L. 94-279, P.L. 99-198) 7 USC 2131 Et. Seq. Implementing Regulations are Published in the Code of Federal Regulations (CFR) Title 9, Chapter 1, Subchapter A, Parts 1, 2, and 3.

Ballantyne, J. C., D. B. Carr, S. deFerranti, et al. 1998. The comparative effects of postoperative analgesic therapies on pulmonary outcome: Cumulative meta-analyses of randomized, controlled trials. *Anesth. Analg.* 86 (3):598–612.

Carbone, E. T., K. E. Lindstrom, S. Diep, and L. Carbone. 2012. Duration of action of sustained-release buprenorphine in 2 strains of mice. *J. Am. Assoc. Lab. Anim. Sci.* 51 (6):815–819.

Chen, F., L. Wang, S. Chen, et al. 2014. Nasal inhalation of butorphanol in combination with ketamine quickly elevates the mechanical pain threshold in the model of chronic constriction injury to the sciatic nerve of rat. *J. Surg. Res.* 186 (1):292–296.

Chum, H. H., K. Jampachairsri, G. P. McKeon, D. C. Yeomans, C. Pacharinsak, and S. A. Felt. 2014. Antinociceptive effects of sustained-release buprenorphine in a model of incisional pain in rats (Rattus Norvegicus). *J. Am. Assoc. Lab. Anim. Sci.* 53 (2):193–197.

Clark, J. A. Jr., P. H. Myers, M. F. Goelz, J. E. Thigpen, and D. B. Forsythe. 1997. Pica behavior associated with buprenorphine administration in the rat. *Lab. Anim. Sci.* 47 (3): 300–303.

Clark, T. S., D. D. Clark, and R. F. Hoyt Jr. 2014. Pharmacokinetic comparison of sustained-release and standard buprenorphine in mice. *J. Am. Assoc. Lab. Anim. Sci.* 53 (4):387–391.

Dahl, J. B. and H. Kchlct. 2011. Preventive analgesia. *Curr. Opin. Anaesthesiol.* 24 (3):331–338.

Dascombe, M. J. 1985. The pharmacology of fever. *Prog. Neurobiol.* 25 (4):327–373.

Fagioli, S., D. Consorti, and F. Pavone. 1990. Age-related cholinergic drug effects on analgesia in two inbred strains of mice. *Brain Res.* 510 (1):150–153.

Flecknell, P. A. 1996. *Laboratory Animal Anaesthesia: An Introduction for Research Workers and Technicians.* 2nd edn. San Diego, CA: Elsevier Academic Press.

Flecknell, P. A., H. E. Orr, J. V. Roughan, and R. Stewart. 1999. Comparison of the effects of oral or subcutaneous carprofen or ketoprofen in rats undergoing laparotomy. *Vet. Rec.* 144 (3):65–67.

Foley, P. L., H. Liang, and A. R. Crichlow. 2011. Evaluation of a sustained-release formulation of buprenorphine for analgesia in rats. *J. Am. Assoc. Lab. Anim. Sci.* 50 (2):198–204.

Foley, P. L., B. D. Ulery, H. M. Kan, et al. 2013. A chitosan thermogel for delivery of ropivacaine in regional musculoskeletal anesthesia. *Biomaterials* 34 (10):2539–2546.

Gades, N. M., P. J. Danneman, S. K. Wixson, and E. A. Tolley. 2000. The magnitude and duration of the analgesic effect of morphine, butorphanol, and buprenorphine in rats and mice. *Contemp. Top. Lab. Anim. Sci.* 39 (2):8–13.

Gillingham, M. B., M. D. Clark, E. M. Dahly, L. A. Krugner-Higby, and D. M. Ney. 2001. A comparison of two opioid analgesics for relief of visceral pain induced by intestinal resection in rats. *Contemp. Top. Lab. Anim. Sci.* 40 (1):21–26.

Goldkuhl, R., J. Hau, and K. S. Abelson. 2010. Effects of voluntarily-ingested buprenorphine on plasma corticosterone levels, body weight, water intake, and behaviour in permanently catheterised rats. *In Vivo* 24 (2):131–135.

Grandin, T. and M. Deesing. 2002. *Distress in Animals: Is it Fear, Pain or Physical Stress?* Manhattan Beach, CA, American Board of Veterinary Practitioners Symposium.

Harper, M. H., R. F. Hickey, T. H. Cromwell, and S. Linwood. 1976. The magnitude and duration of respiratory depression produced by fentanyl and fentanyl plus droperidol in man. *J. Pharmacol. Exp. Ther.* 199 (2):464–468.

Healy, J. R., J. L. Tonkin, S. R. Kamarec, et al. 2014. Evaluation of an improved sustained-release buprenorphine formulation for use in mice. *Am. J. Vet. Res.* 75 (7):619–625.

Interagency Research Animal Committee. 1985. US Government Principles for the Utilization and Care of Vertebrate Animals used in Testing, Research and Training. Federal Register. Vol. May 20, 1985. Washington, DC: Office of Science and Technology Policy.

Izer, J. M., T. L. Whitcomb, and R. P. Wilson. 2014. Atipamezole reverses ketamine-dexmedetomidine anesthesia without altering the antinociceptive effects of butorphanol and buprenorphine in female C57BL/6J mice. *J. Am. Assoc. Lab. Anim. Sci.* 53 (6):675–683.

Jacobson, C. 2000. Adverse effects on growth rates in rats caused by buprenorphine administration. *Lab. Anim.* 34 (2):202–206.

Jirkof, P., A. Tourvieille, P. Cinelli, and M. Arras. 2015. Buprenorphine for pain relief in mice: Repeated injections vs sustained-release depot formulation. *Lab. Anim.* 49 (3):177–187.

Kendall, L. V., D. J. Wegenast, B. J. Smith, et al. 2016. Efficacy of sustained-release buprenorphine in an experimental laparotomy model in female mice. *J. Am. Assoc. Lab. Anim. Sci.* 55 (1):66–73.

Kohn, D. F., T. E. Martin, P. L. Foley, et al. 2007. Public statement: Guidelines for the assessment and management of pain in rodents and rabbits. *J. Am. Assoc. Lab. Anim. Sci.* 46 (2):97–108.

Koutroli, E., P. Alexakos, Z. Kakazanis, et al. 2014. Effects of using the analgesic tramadol in mice undergoing embryo transfer surgery. *Lab. Anim. (NY)* 43 (5):167–172.

Langford, D. J., A. L. Bailey, M. L. Chanda, et al. 2010. Coding of facial expressions of pain in the laboratory mouse. *Nat. Methods* 7 (6):447–449.

Leach, M. C., H. E. Bailey, A. L. Dickinson, J. V. Roughan, and P. A. Flecknell. 2010. A preliminary investigation into the practicality of use and duration of action of slow-release preparations of morphine and hydromorphone in laboratory rats. Lab. Anim. 44 (1):59–65.

Leach, M. C., K. Klaus, A. L. Miller, M. Scotto di Perrotolo, S. G. Sotocinal, and P. A. Flecknell. 2012. The assessment of post-vasectomy pain in mice using behaviour and the mouse grimace scale. *PLoS One* 7 (4):e35656.

Liles, J. H. and P. A. Flecknell. 1992. The use of non-steroidal anti-inflammatory drugs for the relief of pain in laboratory rodents and rabbits. *Lab. Anim.* 26 (4):241–255.

Liles, J. H., P. A. Flecknell, J. Roughan, and I. Cruz-Madorran. 1998. Influence of oral buprenorphine, oral naltrexone or morphine on the effects of laparotomy in the rat. *Lab. Anim.* 32 (2):149–161.

Lutfy, K. and A. Cowan. 2004. Buprenorphine: A unique drug with complex pharmacology. *Curr. Neuropharmacol.* 2 (4):395–402.

McGuire, L., K. Heffner, R. Glaser, et al. 2006. Pain and wound healing in surgical patients. *Ann. Behav. Med.* 31 (2):165–172.

McKeon, G. P., C. Pacharinsak, C. T. Long, et al. 2011. Analgesic effects of tramadol, tramadol-gabapentin, and buprenorphine in an incisional model of pain in rats (Rattus Norvegicus). *J. Am. Assoc. Lab. Anim. Sci.* 50 (2):192–197.

Mao, J. and L. L. Chen. 2000. Gabapentin in pain management. *Anesth. Analg.* 91 (3):680–687.

Matsumiya, L. C., R. E. Sorge, S. G. Sotocinal, et al. 2012. Using the mouse grimace scale to reevaluate the efficacy of postoperative analgesics in laboratory mice. *J. Am. Assoc. Lab. Anim. Sci.* 51 (1):42–49.

Merskey, H. and N. Bogduk, eds. 1994. *Classification of Chronic Pain, IASP Task Force on Taxonomy*. Seattle, WA: International Association for the Study of Pain Press (Also available online at www.iasp-painorg).

Mogil, J. S. and A. L. Bailey. 2010. Sex and gender differences in pain and analgesia. *Prog. Brain Res.* 186:141–157.

Muir, W.W. 2002. Pain and stress. In *Handbook of Veterinary Pain Management*, edited by J. S. Gaynor and W. W. Muir III. Philadelphia, PA: Elsevier Health Sciences.

National Research Council, Institute for Laboratory Animal Research, *Guide for the Care and Use of Laboratory Animals*. 8th ed. National Academies Press, 2011.

Pascoe, P. 2005. The cardiovascular effects of dexmedetomidine given by continuous infusion during isoflurane anesthesia in dogs. *Veterinary Anaesthesia and Analgesia* 32 (4):99.

Perkowski, S.Z. and L.A. Wetmore. 2006. The science and art of analgesia. In *Recent Advances in Veterinary Anesthesia and Analgesia: Companion Animals*, edited by R. D. Gleed and J. W. Ludders. Ithaca, NY: International Veterinary Information Service (IVIS) (www.ivis.org).

Public Health Service (PHS). 1996. *Public Health Service Policy on Humane Care and use of Laboratory Animals*. Washington, DC: US Department of Health and Human Services.

Redwine, K. E. and K. A. Trujillo. 2003. Effects of NMDA receptor antagonists on acute mu-opioid analgesia in the rat. *Pharmacol. Biochem. Behav.* 76 (2):361–372.

Roughan, J. V. and P. A. Flecknell. 2000. Effects of surgery and analgesic administration on spontaneous behaviour in singly housed rats. *Res. Vet. Sci.* 69 (3):283–288.

Roughan, J. V. and P. A. Flecknell. 2001. Behavioural effects of laparotomy and analgesic effects of ketoprofen and carprofen in rats. *Pain* 90 (1–2):65–74.

Roughan, J. V. and P. A. Flecknell. 2004. Behaviour-based assessment of the duration of laparotomy-induced abdominal pain and the analgesic effects of carprofen and buprenorphine in rats. *Behav. Pharmacol.* 15 (7):461–472.

Saper, C. B. and C. D. Breder. 1994. The neurologic basis of fever. *N. Engl. J. Med.* 330 (26):1880–1886.

Schmidt, J. R., L. Krugner-Higby, T. D. Heath, R. Sullivan, and L. J. Smith. 2011. Epidural administration of liposome-encapsulated hydromorphone provides extended analgesia in a rodent model of stifle arthritis. *J. Am. Assoc. Lab. Anim. Sci.* 50 (4):507–512.

Shilo, Y., P. J. Pascoe, D. Cissell, E. G. Johnson, P. H. Kass, and E. R. Wisner. 2010. Ultrasound-guided nerve blocks of the pelvic limb in dogs. *Vet. Anaesth. Analg.* 37 (5):460–470.

Smith, L. J., L. Krugner-Higby, M. Clark, A. Wendland, and T. D. Heath. 2003. A single dose of liposome-encapsulated oxymorphone or morphine provides long-term analgesia in an animal model of neuropathic pain. *Comp. Med.* 53 (3):280–287.

Sotocinal, S. G., R. E. Sorge, A. Zaloum, et al. 2011. The rat grimace scale: A partially automated method for quantifying pain in the laboratory rat via facial expressions. *Mol. Pain* 7:55.

Stasiak, K. L., D. Maul, E. French, P. W. Hellyer, and S. VandeWoude. 2003. Species-specific assessment of pain in laboratory animals. *Contemp. Top. Lab. Anim. Sci.* 42 (4):13–20.

Sternberg, W. F. and J. C. Liebeskind. 1995. The analgesic response to stress: Genetic and gender considerations. *Eur. J. Anaesthesiol. Suppl.* 10:14–17.

Tokuyama, S., M. Inoue, T. Fuchigami, and H. Ueda. 1998. Lack of tolerance in peripheral opioid analgesia in mice. *Life Sci.* 62 (17):1677–1681.

van Loo, P. L., L. A. Everse, M. R. Bernsen, et al. 1997. Analgesics in mice used in cancer research: Reduction of discomfort? *Lab. Anim.* 31 (4):318–325.

Wilson, S. G. and J. S. Mogil. 2001. Measuring pain in the (Knockout) mouse: Big challenges in a small mammal. *Behav. Brain Res.* 125 (1):65–73.

Wright-Williams, S. L., J. P. Courade, C. A. Richardson, J. V. Roughan, and P. A. Flecknell. 2007. Effects of vasectomy surgery and meloxicam treatment on faecal corticosterone levels and behaviour in two strains of laboratory mouse. *Pain* 130 (1–2):108–118.

Zegre Cannon, C., G. E. Kissling, D. R. Goulding, A. P. King-Herbert, and T. Blankenship-Paris. 2011. Analgesic effects of tramadol, carprofen or multimodal analgesia in rats undergoing ventral laparotomy. *Lab. Anim. (NY)* 40 (3):85–93.

euthanasia

C. Tyler Long

contents

Rodent euthanasia has been a topic of debate within the laboratory animal community for the past several decades, with arguments centering around which method of euthanasia is most humane, most efficient, and causes the least amount of impact on research results. Although the subject continues to be actively investigated, most institutions using rodents in research refer to the guidelines published by the AVMA (AVMA) as a reference for rodent and nonrodent euthanasia. For the most current edition (2013) of the AVMA Guidelines for the Euthanasia of Animals, see the following website: https://www.avma.

org/KB/Policies/Documents/euthanasia.pdf (AVMA 2013). Another useful reference utilized by laboratory animal veterinarians is the "Report of the ACLAM Task Force on Rodent Euthanasia," which includes a summary on the biological effects of various euthanasia techniques and can be found at the following website: https://www.aclam.org/Content/files/files/Public/Active/report_rodent_euth.pdf.

8.1 choosing the proper euthanasia method

- When performing rodent euthanasia on healthy or clinically ill animals, there are certain characteristics that the technique chosen should possess. It should
  - Cause minimal pain, distress, and anxiety in the animal
  - Provide rapid loss of consciousness and subsequent loss of cardiac, respiratory, and central nervous system function
  - Be repeatable with similar, effective results
  - Be easy to perform
  - Be safe for personnel to perform
- Additional factors that should be considered prior to selecting which method of euthanasia would be most appropriate for humanely euthanizing rodents includes
  - Operator skill and comfort level
  - Species of animal
  - Number of animals to be euthanized
  - Impact on research results
  - Availability of euthanasia equipment within the facility
  - IACUC approval

8.2 signs of death in rodents

Regardless of the euthanasia method chosen, personnel performing this task must be able to confidently verify death in rodents. Death can be confirmed by thoroughly evaluating the animal and basing their assessment on multiple clinical signs as opposed to a single indicator. (This is especially true of rodents, as certain physical changes may be hard to detect due to their smaller size and positioning within a gas chamber).

- Signs include
  - Prolonged absence of purposeful movements.
  - Absence of reflexive movements, such as response to toe pinch (requires opening of gas chamber, so other signs should be assessed first when using inhalant chamber euthanasia).
  - Respiratory and cardiac arrest. If it is difficult to determine this, then open the chest in smaller rodents to confirm absence of heartbeat—removal of heart can be performed at that time.
  - Loss of bladder function.
  - Cyanosis of mucous membranes.
  - Fixed and dilated pupils.
- If clinical death cannot be assured, a secondary method of euthanasia should be performed prior to the animal regaining consciousness.

8.3 methods of rodent euthanasia

8.3.1 Inhalant Agents

Inhalant agents may be useful when euthanizing multiple rodents at one time or when physical restraint of an animal may be too difficult. The following are general guidelines for rodent euthanasia with inhalant agents:

- Use proper equipment and perform routine scheduled maintenance to ensure repeatable, effective results are achieved each time. All equipment and associated use should conform to local and federal regulations regarding worker safety.
- Agents should always be given to effect and death confirmed prior to removal from anesthetic.
- If the rodent is not dead or if a determination cannot be made, the animal should be immediately reexposed to the inhalant or a secondary physical method should be utilized before the animal regains consciousness.
- When using euthanasia chambers, the space should be large enough to permit each animal to stand on the floor of the chamber with all four feet and have sufficient space to turn

around and perform normal postural movements (Artwohl et al. 2006).

- Rodents should ideally be euthanized in their home cages in order to decrease stress (Artwohl et al. 2006; Anderson 2012; McIntyre et al. 2007). Other steps that can be used to reduce physiologic stress in rodents prior to euthanizing are never performing euthanasia in live animal rooms, euthanizing individual species separately, keeping sick or moribund animals separate from healthy animals, and minimizing noise inside and outside euthanasia chambers.

8.3.1.1 Carbon dioxide

- Carbon dioxide (CO_2) is one of the most commonly utilized methods for humanely euthanizing rodents due to its acceptability with the AVMA Panel on Euthanasia, low cost, and commercial availability. Of all the rodent euthanasia methods, CO_2 arguably best meets the characteristics for choosing the proper euthanasia method listed previously.

- The AVMA Panel on Euthanasia recommends only using CO_2 from compressed gas cylinders (Figure 8.1) in order to best control the inflow of gas into chambers, thus precluding the use of other forms such as chemicals and dry ice (AVMA 2013).

- Although debate still exists on whether to prefill CO_2 chambers prior to placing animals inside (McIntyre et al. 2007; Hawkins et al. 2006; Hewett et al. 1993) the AVMA Panel on Euthanasia currently regards prefilled chambers as unacceptable (AVMA 2013). Placing animals into a chamber that contains room air and gradually introducing CO_2 helps to prevent aversion and the painful sensations seen with high CO_2 concentrations (Artwohl et al. 2006; Anderson 2012).

- A CO_2 fill rate of 20% of chamber volume per minute should be used in order to achieve a lethal concentration (AVMA 2013; Artwohl et al. 2006; Anderson 2012; Hornett and Haynes 1984; Valentine et al. 2012). A 10% V/min rate has been noted to cause less signs of distress in rats (Burkholder et al. 2010). This makes it essential to have a flow meter for measuring CO_2 delivery to the chamber. Below is a simple calculation to

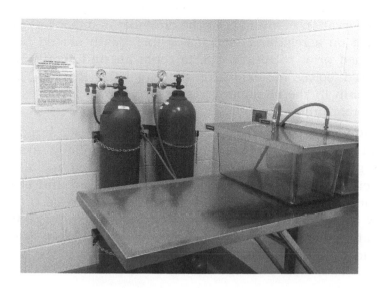

Fig. 8.1 Example of a typical rodent euthanasia chamber with a compressed CO_2 cylinder source. This chamber allows rodents to remain in their home cage and permits visualization of the animals by personnel during the euthanasia procedure.

determine proper CO_2 flow rate to displace 20% of the chamber volume per minute:

Length $(l) \times$ Width $(w) \times$ Height (h) of chamber in inches (in)

$=$ Volume of chamber (in^3)

Volume of chamber $(in^3) \div 61 =$ Volume of chamber in liters (L)

Volume of chamber $(L) \times 20\% =$ flow rate of CO_2 in liters per minute (lpm)

- Adding oxygen to CO_2 can lead to increased time to unconsciousness and distress in animals (AVMA 2013; Artwohl et al. 2006; Anderson 2012). It has also been shown that isoflurane induction and premedicating with acepromazine or midazolam did not decrease distress associated with CO_2 euthanasia and that isoflurane actually increased behavioral signs of stress as compared to 20% V/min CO_2 alone (Valentine et al. 2012).

- Observation of animals in the chamber is important to perform on a regular basis. If possible, death should be confirmed prior to reexposure to room air and should not be based solely on exposure time to CO_2. Rodents should remain in the chamber for several minutes after cessation of breathing.

- Automated CO_2 delivery systems designed for large-scale euthanasia of rodents are now commercially available (Figure 8.2). Advantages of these systems include the facilitation of multistage control of gas input and evacuation, the ability to keep rodents in the home cage, a capacity range from 1 to 160 mouse cages (Euthanex 2014), the fact that they are less labor-intensive, and their use may lead to a reduction in compassion fatigue on the part of

Fig. 8.2 Smartbox™ automated CO_2 system with dual chambers. This two-chamber system can accommodate 40 mice or eight rat cages; larger eight-chamber systems have a capacity of 160 mice or 32 rat cages (Euthanex 2014). Automated CO_2 systems provide a standardized method of controlled gas input and evacuation. (Photo courtesy of Euthanex Corp. dba E-Z Systems, Palmer, PA.)

staff members who have to continually perform euthanasia (Anderson 2012).

8.3.1.2 Gas anesthetics

- Gas anesthetics can be used as a sole means of euthanasia or to cause appropriate unconsciousness to allow for a secondary euthanasia method to be performed in rodents.

- Two common methods for euthanizing rodents with gas anesthetics are the open drop method, where animals are placed in a sealable container with anesthetic-soaked cotton; or by use of an anesthetic vaporizer machine. The latter method may result in prolonged time to death due to intermixing of inhalant gas with oxygen, but delivery is better controlled and contained with a precision vaporizer, thus protecting the user.

- With either method, animals should only be exposed to anesthetic vapors, avoiding the animal coming into contact with the anesthetic agent with the open drop method.

- Isoflurane, halothane, and sevoflurane have rapid onsets of action (Muir et al. 2000c) and levels can be maintained easily with the appropriate vaporizer. Although isoflurane produces a pungent odor that may lead to breath holding and delayed unconsciousness, its rapid induction and commercial availability within the United States make it the preferred gas anesthetic under the current AVMA Guidelines on Euthanasia (AVMA 2013). Sevoflurane has a lower minimum alveolar concentration (MAC) than isoflurane or halothane in most species (Muir et al. 2000b), requiring more drug to achieve loss of consciousness and subsequently death.

- Stutler et al. found that when euthanizing Sprague Dawley rats with halothane, isoflurane, sevoflurane, and enflurane using the open drop method, only euthanasia with halothane and isoflurane overdose resulted in reasonable times to clinical death (5–8 min) (Stutler et al. 2007).

- Due to its slow effect and the possibility of agitation, methoxyflurane is conditionally acceptable for euthanasia in rodents (Wixon and Smiler 1997).

- Diethyl ether poses an explosive hazard within laboratories and is a respiratory irritant, which can induce stress in animals (AVMA 2013; Muir et al. 2000c). Special precautions must be taken when using ether, such as administration in a certified fume hood and disposal of carcasses in explosion-safe freezers.

8.3.1.3 Other inhalants

- Argon is a relatively safe gas for operators to use and can be used in chambers designed for CO_2 euthanasia (Burkholder et al. 2010). Unlike CO_2, a rapid rate of displacement for argon is recommended for euthanasia, as O_2 concentrations of less than 2% need to be achieved rapidly in animals that have been previously sedated or anesthetized (AVMA 2013; Burkholder et al. 2010).

- Recent studies in rats suggest argon as a sole euthanasia agent can be aversive, causing signs such as rearing, gasping, seizure-like activity (Burkholder et al. 2010), tachycardia, and hyperreflexia (Sharp et al. 2006).

- Nitrogen (N_2) is an inert, readily available, and safe gas. Sharp et al. found that in contrast to argon, 100% N_2 did not increase mean arterial pressure and heart rate in rats; however, it was considered ineffective as a euthanasia agent since it was slow to induce unconsciousness and death (Sharp et al. 2006).

- Due to the aversion seen in mice and rats undergoing N_2 and argon euthanasia, the current AVMA Guidelines on Euthanasia recommend the use of other methods for euthanizing laboratory rodents (AVMA 2013).

- At concentrations of 4%–6%, carbon monoxide can cause rapid death by combining with hemoglobin, thus preventing the binding of oxygen (AVMA 2013; AVMA Panel on Euthanasia. AVMA 2001). However, due to the fact that other gaseous agents such as carbon dioxide are readily available and safer for personnel, carbon monoxide is not routinely used for rodent euthanasia. If it is to be used for euthanasia, only compressed cylinders are an acceptable source, and all activities must be performed in well-ventilated areas with constant monitoring of gas levels.

8.3.2 Injectables

- When used properly for euthanasia, injectable agents can cause direct depression of the cerebral cortex and subsequent respiratory and cardiac arrest.

- These drugs have quickest onset when administered intravenously (IV); however, in smaller rodents where IV access is impractical or may cause more stress and anxiety, the intraperitoneal (IP) route may be a better choice.

- Barbituates and dissociative anesthetic combinations can be used as sole agents for euthanasia or used to render rodents unconscious prior to using a secondary method of euthanasia.

- Most injectable agents used for rodent euthanasia are considered controlled substances by the US Drug Enforcement Agency (DEA) and are used extralabel; therefore, these drugs should only be administered under the close supervision of a veterinarian who is currently registered with the US DEA.

8.3.2.1 Barbituates

- Barbituric acid derivatives are widely used in veterinary medicine for euthanasia due to their rapid onset of action, reliable results, low cost, and commercial availability. The most common barbituric acid derivative used in euthanasia of small and large animals is sodium pentobarbital (Muir et al. 2000a).

- Sodium pentobarbital is considered a DEA Schedule II controlled substance, but when it is combined with noncontrolled active ingredients such as phenytoin and lidocaine, it changes the compound from a Class II to a Class III scheduled drug (US Department of Justice Drug Enforcement Administration, Office of Diversion Control 2016; Plumb 1999). These pentobarbital combinations are equally effective as euthanasia agents and are generally easier to obtain and dispense due to their Class III designation.

- Dosages for euthanasia are generally three times that required for anesthesia, which is generally in the range of 150–250 mg/kg IV or IP (Stutler et al. 2007; Hrapkiewicz et al. 1998; Terril and Clemons 1998). If death is not achieved or cannot be verified, subsequent doses should be administered or a secondary method of euthanasia should be performed while the animal is still unconscious.

- Intracardiac injection of pentobarbital may also be used to achieve death in rodents, but it should only be used if the animal is heavily sedated, anesthetized, or unconscious.

8.3.2.2 Dissociative agents

- Although not previously addressed by the AVMA, the 2013 Edition of the Guidelines for the Euthanasia of Animals considers the use of dissociative anesthetic combinations acceptable for the euthanasia of rodents (AVMA 2013).

- Dissociative anesthetics such as ketamine and tiletamine can be combined with α_2-adrenergic receptor agonists or benzodiazepines for surgical anesthesia of rodents (Buitrago et al. 2008; Schoell et al. 2009; Mason 1997). As with pentobarbital, at least three times the anesthetic dose of dissociative combinations should be used for euthanasia and a secondary method employed to ensure death.

- See Chapter 6 ("Special Techniques and Species") for commonly used dissociative agent combination dosages for rodent anesthesia.

- Schoell et al. found that retroorbital administration of 100 μL ketamine (100 mg/mL) and xylazine (100 mg/mL) was a more rapid and efficient method of euthanasia in mice when compared to CO_2 gas and intraperitoneal pentobarbital (Schoell et al. 2009).

8.3.3 Physical Methods

- Physical methods are traditionally utilized as the final step in a two-part process of humane euthanasia where the rodent is first subjected to an inhalant or injectable agent and then a physical method is employed to confirm death.

- Due to the fact that physical methods of euthanasia can be regarded as aesthetically displeasing to staff members, all personnel should not only be highly proficient in the procedure(s) they will be performing but also comfortable with repeatedly performing the procedure(s) in rodents.

- As it may be required by the IACUC, personnel performing a physical method of euthanasia should be prepared to demonstrate their technical proficiency of the procedure in the presence of a trained veterinary staff member.

- In certain instances, inhalants or injectables may impact the results of a study, and with strong scientific justification and prior IACUC approval, cervical dislocation or decapitation may be used as a sole means of euthanasia. Practicing the physical method on heavily anesthetized rodents or carcasses should be performed prior to using the technique as a sole means of euthanasia.

- Controversy exists over whether cervical dislocation and decapitation cause rapid loss of cortical function, leading to minimal pain and distress. The debate centers around the significance of electroencephalogram (EEG) and visual evoked potential (VEPs) recordings conducted in rats and mice (AVMA 2013; Cartner et al. 2007; Mikeska and Klemm 1975; Vanderwolf et al. 1988; Carbone et al. 2012; Bates 2010; Holson 1992). However, the current AVMA guidelines on euthanasia consider these two methods as "acceptable with conditions" (AVMA 2013), meaning that certain conditions must be met to consistently produce humane death, such as proficiency and reproducibility by personnel performing the technique and the use of well-maintained equipment appropriate for the method chosen.
- Table 8.1 summarizes commonly utilized physical methods of rodent euthanasia, including key points on technique, specific requirements, and general comments.

8.4 physiological effects of euthanasia method

- Consideration must be given to how the euthanasia technique(s) chosen will impact study results. This includes how the procedure will affect physiologic parameters and responses of the animal as well as organ and tissue collection.
- To decrease a rodent's physiologic alterations, procedures should be utilized that minimize stress and anxiety to the animal, avoid disruptions to the external environment, and time the euthanasia procedure in accordance with sample analysis and tissue collection.
- A thorough literature search of previous studies can help in choosing which euthanasia method is most compatible with research objectives.
- Euthanasia performed in the presence of live animals can also lead to their undue stress and anxiety and should always be avoided regardless of method.

Table 8.1: Physical Methods of Rodent Euthanasia

Method	Technique	Specific Requirements[a]	Comments
Cervical dislocation	*Manual:* Place thumb and index finger at base of skull. Grab base of tail with opposite hand and retract body forcefully and rapidly at a 30° angle (Figure 8.3a,b) (Cartner et al. 2007; Carbone et al. 2012; AVMA 2013). *Instrument-assisted:* Similar to manual method except instead of thumb and index finger, a sturdy metal rod, closed scissors, hemostat, or metal cage card holder can be placed at base of skull (Figure 8.3c).	• Performed only in mice and rats <200 g. • Animal should be previously subjected to inhalant or injectable euthanasia method unless strong scientific justification exists.	• Quick, efficient, and potentially less stressful than other methods (AVMA 2013; Cartner et al. 2007) • Few equipment demands • Tissues free of chemical or gaseous agents when used alone
Decapitation	Performed in adult rodents with commercial-grade guillotine appropriate for body size (Figure 8.4). Sharp scissors may be used for fetuses and neonates of smaller rodents, or when physical constraints of study do not allow for guillotine use (i.e., ventilator use).	• Equipment must be regularly maintained and kept in proper working condition. • Animal should be previously subjected to inhalant or injectable euthanasia method unless strong scientific justification exists.	• Quick, efficient, and potentially less stressful than other methods (AVMA 2013; Cartner et al. 2007). • Safety concern to personnel if not performed properly • Shown to be less aesthetically pleasing when compared to other methods (Hickman and Johnson 2011) • Period of brief anesthesia prior to procedure can reduce rodent stress (Stutler et al. 2007; Urbanski and Kelley 1991) • Tissues free of chemical or gaseous agents when used alone
Thoracotomy	Under deep sedation or anesthesia, a sharp scalpel blade or pair of scissors is used to transect intercostal muscles, preferably on both sides of chest. Can also transect greater vessels and heart to cause exsanguination in process. Alternatively, the abdominal cavity is opened and diaphragm transected.	• Rodent *must* be heavily sedated or anesthetized prior to performing.	• Not a sole means of euthanasia • Few equipment demands

Exsanguination	Under deep sedation or anesthesia, blood can be collected from cardiac puncture or a major vessel such as the cranial or caudal vena cava. After the required blood is collected into a syringe, the chest is opened and the heart removed or major vessel transected.	• Rodent must be heavily sedated or anesthetized prior to performing.	• Not a sole means of euthanasia. • Few equipment demands. • Larger volumes of blood can be collected for analysis.
Microwave irradiation	Microwave energy focused at the rodent's head causes hyperthermia of the brain, rapidly halting enzyme activity with minimal adverse changes. Because each instrument varies in output and efficiency, instructions for operation should be read carefully before use.	• A machine specifically designed for rodent euthanasia must be used; common kitchen microwaves are unacceptable. • Equipment must be regularly maintained and kept in proper working condition.	• Can lead to death in less than 1 s [AVMA 2013] • Preferred method for immediate fixation of brain metabolites (AVMA 2013). • Specialized equipment limits use to mice and rats and can be costly.

[a] A requirement of all physical methods of euthanasia is that personnel be well trained, proficient, and comfortable in performing the technique. Personnel should be able to demonstrate their abilities in the presence of a trained veterinary staff member prior to performing the procedure unassisted.

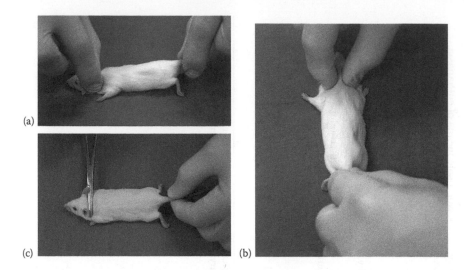

Fig. 8.3 (a,b) Manual cervical dislocation of a mouse previously exposed to CO_2 gas euthanasia. The index finger and thumb are placed at the base of the skull while the other hand grasps the base of the tail and retracts the body at an approximately 30° angle. (c) Cervical dislocation of a mouse previously exposed to CO_2 gas euthanasia using a straight hemostat placed at the base of the skull. The other hand grasps the base of the tail and retracts the body at an approximately 30° angle.

- Discussing how each technique can alter rodent physiology and sample analysis is beyond the scope of this chapter; however, the "Report of the ACLAM Task Force on Rodent Euthanasia" provides a summary on the biological effects of various euthanasia techniques and can be found at the following website: https://www.aclam.org/Content/files/files/Public/Active/report_rodent_euth.pdf.

8.5 fetal and neonatal euthanasia

- There is a limited amount of research available on the topic of fetal and neonatal euthanasia in rodents, however both the ACLAM Task Force on Rodent Euthanasia and the 2013 Edition of the AVMA Guidelines for the Euthanasia of Animals provide general recommendations on the euthanasia of fetal and neonatal rodents.

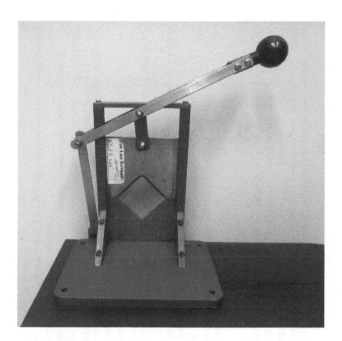

Fig 8.4 Commercial grade guillotine used for decapitation of adult rodents.

- Both references make a specific distinction between precocial and altricial young; with precocial young, such as neonatal guinea pigs, the same euthanasia recommendations are followed as for adults (AVMA 2013; Artwohl et al. 2006).

- Fetuses and neonates have an innate resistance to hypercarbia and hypoxia (AVMA 2013; Pritchett-Corning 2009; Klaunberg et al. 2004; Pritchett et al. 2005), therefore inhalant agents and carbon dioxide may require longer exposure times than in adult rodents. Personnel should be confident in verifying death after using one of these agents in fetuses and neonates or should employ a secondary physical method of euthanasia for confirmation.

- Immersion in liquid nitrogen is not considered a humane form of euthanasia unless fetuses are first rendered unconscious or in altricial neonatal rodents less than 5 days of age (AVMA 2013). Some form of anesthesia should precede immersion or perfusion with chemical fixatives (Artwohl et al. 2006).

- Table 8.2 provides a summary of euthanasia methods for fetal and neonatal rodents.

Table 8.2: Summary of Fetal and Neonatal Rodent Euthanasia

Age	Method of Euthanasia	Comments
Altricial fetuses <15 days in utero Guinea pig fetuses <35 days in utero	Euthanasia of dam or removal of fetus from uterus	• Fetal death in mice at <15 days gestation is equivalent to death of the dam (Pritchett-Corning 2009) • Pain perception is considered unlikely due to lack of neural development (Pritchett-Corning 2009; Artwohl et al. 2006)
Altricial fetuses: In utero day 15–birth Guinea pig fetuses: In utero day 35–birth	Euthanasia of dam	• Best to choose method that minimally disturbs uterus and avoids fetal arousal, such as CO_2 inhalation with secondary physical method of euthanasia (Artwohl et al. 2006; Pritchett et al. 2005) • Pentobarbital IV can be given to avoid accidental in utero injection with IP route, but may require higher dosages in gravid females (Pritchett-Corning 2009).
	Euthanasia of fetuses (If dam not euthanized): A) Inhalants	• Longer exposure times required • Follow up with secondary physical method to confirm death
	B) Injectables	• Overdose of barbituate or dissociative anesthetic combinations administered IP followed by physical method to ensure death.
	C) Decapitation/cervical dislocation	• Decapitation performed with sharp scissors. Use same general guidelines as for adult rodents. • A barrier such as gauze or latex glove should be placed between ice source and fetus. • Follow up with secondary physical method to confirm death.
	D) Hypothermia for anesthesia	
Altricial neonates (For precocial neonates, follow adult euthanasia recommendations)	Inhalants	• Longer exposure times required. • Follow up with secondary physical method to confirm death. • In mice and rats, time needed to achieve death decreased steadily as pups aged (Pritchett et al. 2005; Klaunberg et al. 2004; Pritchett-Corning 2009) • 5% Halothane was ineffective for euthanasia of mouse pups (Pritchett-Corning 2009)
	Injectables	• Overdose of barbiturate or dissociative anesthetic combinations administered IP followed by physical method to ensure death. • Pentobarbital (800 mg/kg IP) caused death in approximately 6 min in 8–14 day old mouse pups (Pritchett-Corning 2009)
	Decapitation/cervical Dislocation	• Decapitation performed with sharp scissors or with guillotine for larger neonates. Use same general guidelines as for adult rodents.
	Hypothermia for anesthesia	• Only in neonates less than 7 days old (AVMA 2013; Artwohl et al. 2006) • A barrier such as gauze or latex glove should be placed between ice source and neonate. • Follow up with secondary physical method to confirm death.

references

American Veterinary Medical Association. 2013. AVMA guidelines for the euthanasia of animals. https://www.avma.org/KB/Policies/Documents/euthanasia.pdf (accessed January 7, 2016).

Anderson, L. 2012. Considerations for implementing a facility-wide policy on carbon dioxide euthanasia for laboratory rodents. http://www.alnmag.com/articles/2012/04/considerations-implementing-facility-wide-policy-carbon-dioxide-euthanasia-laboratory-rodents (accessed January 6, 2016).

Artwohl, J., P. Brown, B. Corning, S. Stein, and ACLAM Task Force. 2006. Report of the ACLAM task force on rodent euthanasia. *JAALAS* 45(1):98–105.

AVMA Panel on Euthanasia. AVMA. 2001. 2000 Report of the AVMA panel on euthanasia. *J. Am. Vet. Med. Assoc.* 218(5):669–696.

Bates, G. 2010. Humane issues surrounding decapitation reconsidered. *J. Am. Vet. Med. Assoc.* 237(9):1024–1026.

Buitrago, S., T. E. Martin, J. Tetens-Woodring, A. Belicha-Villanueva, and G. E. Wilding. 2008. Safety and efficacy of various combinations of injectable anesthetics in BALB/c mice. *JAALAS* 47(1):11–17.

Burkholder, T. H., L. Niel, J. L. Weed, L. R. Brinster, J. D. Bacher, and C. J. Foltz. 2010. Comparison of carbon dioxide and argon euthanasia: Effects on behavior, heart rate, and respiratory lesions in rats. *JAALAS* 49(4):448–453.

Carbone, L., E. T. Carbone, E. M. Yi, et al. 2012. Assessing cervical dislocation as a humane euthanasia method in mice. *JAALAS* 51(3):352–356.

Cartner, S. C., S. C. Barlow, and T. J. Ness. 2007. Loss of cortical function in mice after decapitation, cervical dislocation, potassium chloride injection, and CO_2 inhalation. *Comp. Med.* 57(6):570–573.

Euthanex. 2014. Euthanex with smartbox auto CO_2 system. http://www.euthanex.com/smartbox/ (accessed January 6, 2016).

Hawkins, P., L. Playle, H. Golledge, et al. 2006. Newcastle consensus meeting on carbon dioxide euthanasia of laboratory animals. https://www.nc3rs.org.uk/sites/default/files/documents/Events/First%20Newcastle%20consensus%20meeting%20report.pdf (accessed January 6, 2016).

Hewett, T. A., M. S. Kovacs, J. E. Artwohl, and B. T. Bennett. 1993. A comparison of euthanasia methods in rats, using carbon dioxide in prefilled and fixed flow rate filled chambers. *Lab. Anim. Sci.* 43(6):579–582.

Hickman D. L. and S. W. Johnson. 2011. Evaluation of aesthetics of physical methods of euthanasia of anesthetized rats. *JAALAS* 50(5):695–701.

Holson, R. R. 1992. Euthanasia by decapitation: Evidence that this technique produces prompt, painless unconsciousness in laboratory rodents. *Neurotoxicol. Teratol.* 14(4):253–257.

Hornett, T. D. and A. P. Haynes. 1984. Comparison of carbon dioxide/air mixture and nitrogen/air mixture for the euthanasia of rodents: Design of a system for inhalation euthanasia. *Animal Tech.* 35:93–99.

Hrapkiewicz, K., L. Medina, and D. D. Holmes. 1998. Hamster euthanasia. In *Clinical Laboratory Animal Medicine, an Introduction.* 2nd edition, 82. Ames, IA: Iowa State University Press.

Klaunberg, B. A., J. O'Malley, T. Clark, and J. A. Davis. 2004. Euthanasia of mouse fetuses and neonates. *Contemp. Top. Lab. Anim. Sci.* 43(5):29–34.

Mason, D. E. 1997. Anesthesia, analgesia, and sedation for small mammals. In *Ferrets, Rabbits, and Rodent Clinical Medicine and Surgery,* 388. Philadelphia, PA: W.B. Saunders.

McIntyre, A. R., R. A. Drummond, E. R. Riedel, and N. S. Lipman. 2007. Automated mouse euthanasia in an individually ventilated caging system: System development and assessment. *JAALAS* 46(2):65–73.

Mikeska, J. A. and W. R. Klemm. 1975. EEG evaluation of humaneness of asphyxia and decapitation euthanasia of the laboratory rat. *Lab. Anim. Sci.* 25(2):175–179.

Muir, W. M., J. A. Hubbell, R. T. Skarda, and R. M. Bednarski. 2000a. Euthanasia. In *Handbook of Veterinary Anesthesia,* ed. J. Schrefer. 3rd edition, 496–501. St. Louis, MO: Mosby.

Muir, W. M., J. A. Hubbell, R. T. Skarda, and R. M. Bednarski. 2000b. Inhalation anesthesia. In *Handbook of Veterinary Anesthesia,* ed. J. Schrefer. 3rd edition, 154–163. St. Louis, MO: Mosby.

Muir, W. M., J. A. Hubbell, R. T. Skarda, and R. M. Bednarski. 2000c. Pharmacology of inhalation anesthetic drugs. In *Handbook of Veterinary Anesthesia,* ed. J. Schrefer. 3rd edition, 164–181. St. Louis, MO: Mosby.

Plumb, D. C. 1999. Euthanasia agents containing pentobarbital. In *Veterinary Drug Handbook*. 3rd edition, 305. Ames, IA: Iowa State University Press.

Pritchett, K., D. Corrow, J. Stockwell, and A. Smith. 2005. Euthanasia of neonatal mice with carbon dioxide. *Comp. Med.* 55(3):275–281.

Pritchett-Corning, K. R. 2009. Euthanasia of neonatal rats with carbon dioxide. *JAALAS* 48(1):23–27.

Schoell, A. R., B. R. Heyde, D. E. Weir, P. C. Chiang, Y. Hu, and D. K. Tung. 2009. Euthanasia method for mice in rapid time-course pulmonary pharmacokinetic studies. *JAALAS* 48(5):506–511.

Sharp, J., T. Azar, and D. Lawson. 2006. Comparison of carbon dioxide, argon, and nitrogen for inducing unconsciousness or euthanasia of rats. *JAALAS* 45(2):21–25.

Stutler, S. A., E. W. Johnson, K. R. Still, D. J. Schaeffer, R. A. Hess, and D. P. Arfsten. 2007. Effect of method of euthanasia on sperm motility of mature Sprague-Dawley rats. *JAALAS* 46(2):13–20.

Terril, L. A. and D. J. Clemons. 1998. Euthanasia. In *The Laboratory Guinea Pig*, ed. M. A. Suckow, 94–95. Boca Raton, FL: CRC Press.

Urbanski, H. F and S. T. Kelly. 1991. Sedation by exposure to a gaseous carbon dioxide-oxygen mixture: Application to studies involving small laboratory animal species. *Lab. Anim. Sci.* 41(1):80–82.

US Department of Justice Drug Enforcement Administration, Office of Diversion Control. 2016. List of Controlled Substances. http://www.deadiversion.usdoj.gov/schedules/index.html (accessed January 6, 2016).

Valentine, H., W. O. Williams, and K. J. Maurer. 2012. Sedation or inhalant anesthesia before euthanasia with CO_2 does not reduce behavioral or physiologic signs of pain and stress in mice. *JAALAS* 51(1):50–57.

Vanderwolf, C. H., G. Buzsaki, D. P. Cain, R. K. Cooley, and B. Robertson. 1988. Neocortical and hippocampal electrical activity following decapitation in the rat. *Brain Res.* 451(1–2):340–344.

Wixon, S. K. and K. L. Smiler. 1997. Anesthesia and analgesia in rodents. In *Anesthesia and Analgesia in Laboratory Animals*, eds. D. F. Kohn, S. K. Wixson, W. J. White, and G. J. Benson, 165–203. San Diego, CA: Academic Press.

regulatory management of rodent anesthesia

Patrick Sharp

contents

Rodent regulatory management is more perplexing than that of other animals because of the legal standing seen in rodents. This is confounded by the fact that more than 90% of animals used in biomedical research in the United States are rodents. The US approach to the regulatory management of biomedical research animals is primarily self-monitoring, but inspection reports are available from the United States Department of Agriculture (USDA) for "covered" species used in biomedical research.

While some rodents (e.g., guinea pigs, hamsters, gerbils) are "covered" by the law (US *Animal Welfare Act*, AWA) and regulations (*Animal Welfare Regulations*, AWR) others are not (e.g., research purpose–bred rats and mice). Specifically, the AWA states, "The term 'animal' means any live or dead... guinea pig, hamster...or such other warm-blooded animal, as the Secretary may determine is being used, or is intended for use, for research, testing, experimentation, or exhibition purposes, or as a pet; but such term excludes (1) birds, rats of the genus *Rattus*, and mice of the genus *Mus*, bred for use in research...," and the AWR carry virtually identical language. To avoid the differences and confusion that may occur between those species that are "covered" or not, many institutions choose to "cover" all species used in biomedical research.

Research staff should strive to utilize the key elements of regulatory management established in the rules and regulations below. Building strong, good-natured, professional relationships with the veterinary staff (Attending Veterinarian), Institutional Animal Care and Use Committee (IACUC), and the institutional administration (Institutional Official) is critical. Cultivating and maintaining constructive relationships with each group will facilitate institutional and individual research. Proactively encountering these groups is essential during recruitment to an institution, especially if the research is considered even remotely controversial or complex. These three groups can work with researchers to facilitate the research process to the best of their abilities. Furthermore, waiting until a problem develops or festers may not yield an appropriate constructive relationship.

9.1 virtues, values, and ethics

Ethical and responsible animal use in biomedical research is essential to ensure the public's trust in the activities undertaken. Scientists must be aware that public acceptance of animal use in biomedical research is a changing dynamic endeavor. Driving the use of animals are three interrelated but independent concepts: virtues, values, and ethics. Expediency can lead to research-related problems due to misunderstanding of the research aims and can foul the research integrity of an institution, individual, or scope of research.

9.1.1 The 3Rs

- The 3Rs originated from Russell and Burch's *The Principles of Humane Experimental Technique* (1959): replacement, reduction, and refinement. Briefly, replacement is the use of *in vitro* or *in silico* models instead of live animals; reduction is the use of the minimum number of animals; and refinement is, among other things, the use of analgesia and anesthesia and the minimal use of injurious methods and procedures.

- Research protocols, especially those involving anything more than momentary pain or distress, will require researchers to demonstrate how they are applying the 3Rs and seeking alternatives.

9.1.2 Replacement

- Using *in vitro* or *in silico* models rather than live animals; few of these models currently exist.

- Biologic models are elaborate with multiple, integrated organ systems and only rarely permit independent investigation.

- Animal cell cultures are reliant upon animals for cell culture origin and/or feeding the cell cultures.

- Consideration should be given to phylogenically lower species when possible, including invertebrates such as *C. elegans* and fruit flies.

- The Koken Rat® is available and may be useful for training purposes.
- "By considering opportunities for refinement, the use of appropriate nonanimal alternatives, and the use of fewer animals, both the institution and the principal investigator (PI) can begin to address their shared obligations for humane animal care and use." (*Guide for the Care and Use of Laboratory Animals* [*Guide*], p. 27).

9.1.3 Reduction

- Reducing animal numbers without sacrificing the scientific quality of the research
- Animal reduction strategies include
  - Prudent experimental design, including discussions with a statistician and veterinary pathologist
  - Valuing and maintaining well-trained research, husbandry, and veterinary teams aware of their limitations
    - Value and support their continuing education to meet ever-changing research demands
    - Support to strategically import expertise when confronting new research challenges (e.g., veterinary surgeon, veterinary anesthesiologist)
  - Controlling research interfering experimental variables, including
    - Using genetically homogeneous animals, where feasible
    - Evaluating dietary components for research-confounding problems (e.g., phytoestrogens)
    - Using animals free of deleterious infectious and commensal agents
  - Minimizing the use of control animals
  - Using optimal or targeted animal models

9.1.4 Refinement

9.1.4.1 Refine research procedures and techniques by

- Safer, better controlled anesthetic regimens
- Multimodal analgesia/anesthesia regiments
- Pain scoring

- Pilot studies
- Better surgical techniques training using models or cadavers before using live animals.
- Using facial vein or saphenous vein blood collection techniques rather than retroorbital collection techniques
- Using hypothermia rather than death as an endpoint or LD50 studies
- Using alternatives to tail tipping for animal genotyping

9.1.4.2 Refine husbandry techniques

- Group housing
- Environmental enrichment

9.1.5 United States Department of Agriculture

The USDA has developed pain and distress categories; many institutions (domestic and foreign) apply these pain categories to rats and mice:

- Pain category C: Animals that are subject to procedures that cause no pain or distress or only momentary or slight pain or distress and do not require the use of pain-relieving drugs.
- Pain category D: Animals subjected to potentially painful or stressful procedures for which they receive appropriate anesthetics, analgesics, and/or tranquilizer drugs.
- Pain category E: Animals subjected to potentially painful or stressful procedures that are *not* relieved with anesthetics, analgesics, and/or tranquilizer drugs. Withholding anesthesia/analgesia must be scientifically justified in writing and approved by the IACUC.

9.2 legislative and nonlegislative standards overview

Those performing research must be aware of the legislative and nonlegislative standards described below at a minimum. Each institution provides various compliance mechanisms for these and other regulations impacting research, teaching, and testing. Failure to follow these mechanisms can seriously and negatively impact institutional and individual consequences. The regulatory landscape, like the scientific landscape, is constantly changing. Cultivating a strong, professional, and collegial relationship between and among research staff, veterinary staff, IACUC, and administration creates a strong foundation for regulatory compliance benefiting both individual and institutional research programs.

Those performing research must be acutely aware of the procedures, compounds, and so on outlined in their IACUC-approved protocol and must consult the IACUC and the Attending Veterinarian before making protocol-related changes.

9.2.1 Federal Agencies/Standards

9.2.1.1 United States Department of Agriculture

9.2.1.1.1 Animal Welfare Act

- Promulgated and enforced by the USDA.
- Provides specifically for the USDA's Animal and Plant Health Inspection Service (APHIS) to administer and enforce the AWA.
- Among other matters, it is a law that sets out general expectations for the humane care and treatment of research animals, including some rodents.
- Many institutions expand the legal aspects of the AWA to all rats and mice used in research.
- http://www.aphis.usda.gov/animal_welfare/downloads/ Animal%20Care%20Blue%20Book%20 %202013%20-%20 FINAL.pdf

9.2.1.1.2 Animal welfare regulations

- The USDA's specific expectations for, among other matters outlined in the AWA, the humane care and use of research animals, including some rodents.
- http://www.aphis.usda.gov/animal_welfare/downloads/ Animal%20Care%20Blue%20Book%20-%202013%20-%20 FINAL.pdf

9.2.1.1.3 Animal care resource guide policies

- Promulgated and enforced by the USDA
- Elucidates the AWA's intent by providing the USDA's interpretations of select matters pertaining to the AWA and AWR
- http://www.aphis.usda.gov/animal_welfare/downloads/ Animal%20Care%20Policy%20Manual.pdf

9.2.1.2 Office of Laboratory Animal Welfare (OLAW)

- "OLAW provides guidance and interpretation of the Public Health Service (PHS) Policy on Humane Care and Use of Laboratory Animals, supports educational programs, and monitors compliance with the Policy by Assured institutions and PHS funding components to ensure the humane care and use of animals in PHS-supported research, testing, and training, thereby contributing to the quality of PHS-supported activities."

- Negotiates the written assurance outlines in *PHS Policy* §IV. A.

- Assurances are negotiated with both foreign and domestic institutions receiving PHS funding.

- Each institution will provide an annual report according to §IV. F.

- http://grants.nih.gov/grants/olaw/olaw.htm

9.2.1.2.1 *US government principles for the utilization and care of vertebrate animals used in testing, research, and training*

- Promulgated and enforced by the US National Institutes of Health's (NIH's) OLAW

- http://grants.nih.gov/grants/olaw/references/phspol. htm#USGovPrinciples

9.2.1.2.2 *Health research extension act*

- Promulgated and enforced by the NIH's OLAW

- Public Law 99-158, 1985

- http://grants.nih.gov/grants/olaw/references/hrea1985.htm

9.2.1.2.3 *Public health service policy on humane care and use of laboratory animals*

- Promulgated and enforced by the NIH's OLAW through the Department of Health and Human Services.

- §IV. A states, "No activity involving animals may be conducted or supported by the PHS until the institution conducting the activity has provided a written Assurance acceptable to the PHS, setting forth compliance with the Policy."

- Assurance statements are required for both domestic and foreign institutions receiving PHS funds.

9.2.1.3 US Drug Enforcement Administration (DEA)

- Falls under the US Department of Justice.
- Enforces the *Controlled Substances Act, Code of Federal Regulations Title 21 Chapter 13.*
- http://www.deadiversion.usdoj.gov/21cfr/21usc/index.html
- Applicants for a controlled substances permit must be compliant with federal, state, and local laws.
- Many research institutions develop their own policies pertaining to controlled substances. These frequently include the IACUC.

9.2.1.4 US Food and Drug Administration (FDA)

- Falls under the Department of Health and Human Services.
- Enforces the *Good Laboratory Practices*: *Code of Federal Regulations Title 21, Chapter 1, Part 58*

9.2.2 State/Local Regulations

In addition to federal regulations, some states and localities may have regulations that require permits or prohibit the entry of some rodent species. The scope of the regulations may not clearly distinguish those rodents used in biomedical research; furthermore, when exemptions are given for research purposes, the permits may only be for "white mice" and "white rats." Permits may be required for some rodent species; again, exemptions may not be made for those animals used exclusively in biomedical research. To secure a permit for some species, a facility inspection and/or additional facility design requirements may be required.

Perhaps the best known example of this regulation is California's Detrimental Species Legislation *California Department of Fish and Wildlife* (§671. Importation, Transportation and Possession of Live Restricted Animals). The National Association for Biomedical Research (NABR) maintains a listing on their website of various US state laws and regulations.

9.2.2.1 Environmental health and safety (EHS) and occupational health and safety (OHS)

- While the AWA is silent on EHS and OHS issues, the *Guide for the Care and Use of Laboratory Animals* (*Guide*) dedicates a significant portion of the document to matters impacting both animal and human health and safety.

- Hazards in biomedical research are regulated from federal, state, and local agencies. IACUCs work closely with institutional EHS and OHS professionals to facilitate regulatory compliance.
- The following document may be of assistance:
  - *Occupational Health and Safety in the Care and Use of Research Animals* (NRC 1997)

9.2.3 Nonlegislative Standards

9.2.3.1 Guide for the care and use of laboratory animals

- Published by the Institute of Laboratory Animal Research (ILAR).
- Adopted by government and nongovernment organizations (e.g., NIH, Association for the Assessment and Accreditation of Laboratory Animal Care—International [AAALAC]); the standard was "updated" in 2011.
- Serves as one of AAALAC's three primary standards for evaluating laboratory animal care and use programs.1.
- http://www.aaalac.org/resources/Guide_2011.pdf

9.2.3.2 American Veterinary Medical Association guidelines on euthanasia (See Chapter 8)

- Periodically rewritten by the American Veterinary Medical Association (AVMA).
- Provides humane, scientifically-based euthanasia methods.
- Previously referred to as the *AVMA Panel on Euthanasia.*
- https://www.avma.org/KB/Policies/Documents/euthanasia.pdf

9.2.3.3 Guidelines of the American Society of Mammalogists for the use of wild mammals in research

- Periodically rewritten by the American Society of Mammalogists and includes rodents
- http://www.mammalsociety.org/uploads/Sikes%20et%20al%202011.pdf

9.2.3.4 Guidelines for the care and use of mammals in neuroscience and behavioral research

- Periodically rewritten; published by ILAR.

- Previously entitled *Preparation and Maintenance of Higher Mammals during Neuroscience Experiments.*

- Also known previously as the "Red Book."

- The current version now includes rodents.

- http://grants.nih.gov/grants/OLAW/National_Academies_Guidelines_for_Use_and_Care.pdf

9.2.3.5 Institutional policies and standard operating procedures

The research "mix" conducted at each institution varies and has different drivers. The varied research mix is captured in the IACUC and the institutional research programs it oversees as it strives to meet the institution's regulatory requirements. Each IACUC's goal should be to work with researchers to secure IACUC approval of the various research proposals. To facilitate the process, it is suggested that IACUCs develop various policies to assist the researcher in writing the research proposals.

First and foremost, an institution must answer the question, "Who can be a PI?" While some institutions feel it should *only* be a research faculty member, they should keep in mind that developing, writing, and managing an IACUC protocol is a learning experience consistent with undergraduate, graduate, and postdoctoral work, not just with being a faculty member. This specific question is frequently more than just an IACUC matter and also needs the attention of the institutional official (IO).

Other institutional policies for the care and use of animals fall under the general responsibility of the IACUC; these policies can be further divided into various categories including the following:

9.2.3.5.1 General

- Veterinary responsibilities/authority
- Investigator responsibilities
- IACUC
  - Protocol review procedures
  - Investigating noncompliance

9.2.3.5.2 *Animal care, husbandry, management, and housing*

- Acclimation and quarantine
- Special diet
- Environmental enrichment

9.2.3.5.3 *Procedure and procedure-related*

- Surgical procedures:
  - Survival surgery guidelines
  - Nonsurvival surgery guidelines
  - Anesthesia analgesia and tranquilizer guidelines
  - Maintaining chronic implants

9.2.3.5.4 *Nonsurgical procedures (may still necessitate using anesthesia)*

- Blood and tissue collection
- Monoclonal antibody production

9.2.3.5.5 *Special considerations*

- Restricted food and water access
- Prolonged physical restraint

9.2.3.5.6 *Animal transportation*

- Intra-institutional
- Extra-institutional

9.2.3.5.7 *Euthanasia*

- Studies on death as an endpoint
- Guidelines for euthanasia by physical methods
- Euthanasia of feti and neonates

9.2.3.5.8 *Other*

- Occupational health issues
- Tissue sharing

Some institutions use institutional standard operating procedures (SOPs), which are included in a research protocol rather than necessitating (re)writing the procedure. While these institutional SOPs may work for some small organizations, when multiple users are using a given SOP, they must be cognizant that if the "owner" of the SOP makes substantive changes, by default all individuals using the same SOP could find themselves in a situation where they are performing work outside the SOP's scope. Conversely, a PI who wants to change an item in the SOP is in effect creating a new SOP. Institutional SOPs can add unnecessary complications to the research program. Using the "boiler plate" language within the SOPs may have more benefit than developing, coordinating, and accurately using institutional SOPs.

9.3 Association for Assessment and Accreditation of Laboratory Animal Care—International (AAALAC—International)

- Nonprofit organization providing voluntary, peer-reviewed accreditation of institutions using animals in research, teaching, and testing
- Established in 1965
- Provides a confidential accreditation
- Perceived as the "gold standard of laboratory animal care"
- Uses three regulatory standards:
  - The *Guide for the Care and Use of Laboratory Animals* (*Guide*)
  - The *Guide for the Care and Use of Agricultural Animals in Research and Teaching* (*Ag Guide*, not applicable to rodents)
  - ETS 123, Appendices A and B
- US institutions: Guide and other pertinent regulations apply.
- Outside the US:
  - National regulations apply.
  - In the case of a foreign regulatory "void" (e.g., occupational health and safety), the *Guide* or associated regulation applies.
  - In the case of a regulatory conflict occurs between a nation's regulations and the *Guide*, the stricter of the two applies.

9.4 legislative and nonlegislative standard highlights

9.4.1 Training

9.4.1.1 The Guide

- The *Guide* (pp. 15–17) states, "All personnel involved with the care and use of animals must be adequately educated, trained, and/or qualified in basic principles of laboratory animal science to help ensure high-quality science and animal well-being."

- "the IACUC is responsible for providing oversight and for evaluating the effectiveness of the training program." (p. 15)

- It is important to document each training event; if it is not documented or documented improperly, then the event didn't happen.

- The *Guide* specifically highlights the training expectations of veterinarians and professional staff, animal care personnel, the research team, and the IACUC.

- The *Guide* underscores the importance of training and its role in the occupational health/safety and investigating animal welfare concerns of *all* personnel (pp. 17–24), "including personnel such as custodial, maintenance, and administrative staff, who are farther removed from the animal use."

- IACUC protocol review includes training adequacy (p. 26), which includes training personnel accountable for assessing and recognizing humane endpoints and the required action when those endpoints are reached (pp. 27–28).

- Training should incorporate into the disaster plan actions to be taken in the event of unplanned events (e.g., earthquakes) and disaster/emergency training (e.g., scheduled fire alarms).

- Training should include the identification of "adverse and abnormal behaviors" for the various species.

- Instruction and training for intra-institutional animal movement is critical to ensure that the animals are transported from the animal holding area to the procedure area and back to an appropriate postprocedure holding area. The postprocedure housing area may be a "return room" and not the animal's original animal holding room due to various factors; it is important to thoroughly understand animal transportation as it pertains to one's research proposal.

- Daily observations (e.g., once daily 7 days a week) of animals undergoing procedures is a requirement for good animal care and an essential component of a sound scientific method. Observations should note an animal's behavior, posture, food/water consumption, and body weight, and should include an evaluation of the procedure area/site along with appropriate entries into the animal's clinical record; the IACUC should clearly outline a research proposal's observation expectations. The clinical record must include the administration of research-related compound(s), analgesics, and/or antibiotics. Depending on the scope and other factors associated with the research paradigm, it may be prudent (or required) to evaluate an animal multiple times per day, including documented observation after hours and on weekends and/or holidays. With current security systems, including cameras, it is easy to determine if the person(s) responsible for animal observations actually entered the facility and/or a given room.

9.4.2 Animal Welfare Regulations

9.4.2.1 §2.31 Institutional Animal Care and Use Committee (IACUC)

- §2.31(c)(4) states, "Make recommendations to the institutional official regarding any aspect of the research facility's animal program, facilities, or personnel training."
- §2.31(d)(1)(viii) states, "Personnel conducting procedures on the species being maintained or studied will be appropriately qualified and trained in those procedures."

9.4.2.2 §2.32 Personnel qualifications

- §2.32(a) "It shall be the responsibility of the research facility to ensure that all scientists, research technicians, animal technicians, and other personnel involved in animal care, treatment, and use are qualified to perform their duties. This responsibility shall be fulfilled in part through the provision of training and instruction to those personnel."
- §2.32(b) "Training and instruction shall be made available, and the qualifications of personnel reviewed, with sufficient

frequency to fulfill the research facility's responsibilities under this section and §2.31."

- §2.32(c) "Training and instruction of personnel must include guidance in at least the following areas:
  - §2.32(c)(1) Methods of animal maintenance and experimentation, including:
    - The basic needs of each species of animal.
    - Proper handling and care for the various species of animals used by the facility.
    - Proper pre-procedural and post procedural care of animals.
    - Aseptic surgical methods and procedures."
  - §2.32(c)(2) "The concept, availability, and use of research or testing methods that limit the use of animals or minimize animal distress."
  - §2.32(c)(3) "Proper use of anesthetics, analgesics, and tranquilizers for any species of animals used by the facility."
  - §2.32(c)(4) "Methods whereby deficiencies in animal care and treatment are reported, including deficiencies in animal care and treatment reported by any employee of the facility. No facility employee, Committee member, or laboratory personnel shall be discriminated against or be subject to any reprisal for reporting violations of any regulation or standards under the Act."
  - §2.32(c)(5) "Utilization of services (e.g., National Agricultural Library, National Library of Medicine) available to provide information:
    - On appropriate methods of animal care and use.
    - On alternatives to the use of live animals in research.
    - That could prevent unintended and unnecessary duplication of research involving animals.
    - Regarding the intent and requirements of the Act."

9.4.3 Animal Care Resource Guide Policies

Policy #3, Veterinary Care: Elucidates the expectations for various topics germane to this chapter, including

- *Expired medical materials*: The use of expired medical materials such as drugs, fluids, or sutures on regulated animals is not considered to be acceptable veterinary practice and does not constitute adequate veterinary care as required by the regulations promulgated under the Animal Welfare Act....The facility must either dispose of all such materials or segregate them in an appropriately labeled, physically separate location from non-expired medical materials....Proper anesthesia, analgesia, and euthanasia are required for all such procedures. Drugs administered to relieve pain or distress and emergency drugs must not be used beyond their expiration date. Facilities allowing the use of expired medical materials in acute terminal procedures should have a policy covering the use of such materials and/or require investigators to describe in their animal activity proposals the intended use of expired materials.

- *Pharmaceutical-grade compounds in research*: Investigators are expected to use pharmaceutical-grade medications whenever they are available, even in acute procedures. Non-pharmaceutical-grade chemical compounds should only be used in regulated animals after specific review and approval by the IACUC for reasons such as scientific necessity or the nonavailability of an acceptable veterinary or human pharmaceutical-grade product. Cost savings alone are not an adequate justification for using non-pharmaceutical-grade compounds in regulated animals.

- *Surgery*

- *Pre- and postprocedural care*

- *Program of veterinary care*

- *Health records*

- *Euthanasia*

Policy #12, Consideration of Alternatives to Painful/Distressful Procedures: "The Animal Welfare Information Center (AWIC) is an information service of the National Agricultural Library specifically established to provide information about alternatives. AWIC offers expertise in formulation of the search strategy and selection of terminology and databases, access to unique databases, on- and off-site training of institute personnel in conducting effective alternatives searches, and is able to perform no-cost or low-cost electronic database searches. AWIC can be contacted at (301) 504-6212, via E-mail at awic@nal.usda.gov, or via its web site at http://www.awic.nal. usda.gov. Other excellent resources for assistance with alternative searches are available and may be equally acceptable."

Policy #15, Institutional official and IACUC Membership: "IACUC members must be qualified to assess the research facility's animal program, facilities and procedures. The research facility is responsible for ensuring their qualification, and this responsibility is filled in part through the provision of training and instruction. For example, IACUC members should be trained in understanding the Animal Welfare Act, protocol review, and facility inspections."

9.4.4 US Government Principles for the Utilization and Care of Animals Used in Testing, Research, and Training

- "Whenever U.S. Government agencies develop requirements for testing, research, or training procedures involving the use of vertebrate animals, the following principles shall be considered; and whenever these agencies actually perform or sponsor such procedures, the responsible IO shall ensure that these [9] principles are adhered to."

- "VIII. Investigators and other personnel shall be appropriately qualified and experienced for conducting procedures on living animals. Adequate arrangements shall be made for their in-service training, including the proper and humane care and use of laboratory animals."

9.4.5 Training Sources

Various organizations can assist with institutional training. These organizations include

- American Association for Laboratory Animal Science (AALAS)
- Laboratory Animal Welfare and Training Exchange (LAWTE)
- Animal Welfare Information Center (AWIC)
- Academy of Surgical Research

Other resources include

- *Education and Training in the Care and Use of Laboratory Animals: A Guide for Developing Institutional Programs* (1991)
- Local AALAS branch meetings

9.4.6 Key Concepts

- Personnel training is an absolutely critical component that directly reflects on an institution's animal care and use program and culture of care.
- The IACUC's unaffiliated members and nonscientists must familiarize themselves with the training program directly in a "hands-on" manner as the "public conscience" of the animal care and use program.
- IACUC members must take an active role in ensuring that training programs meet current and evolving institutional needs and play a part in minimizing institutional deficiencies.
- "Grandfather clauses" are poor substitutes for training programs and suggest kowtowing to select institutional groups and/or weakness with the IO to appropriately support the IACUC in whole or part.
- Language and cultural barriers

9.5 procedures

The IACUC must inspect all animal care and use facilities, including surgical areas, every 6 months. Institutions may wish to provide centralized/dedicated rodent procedure facilities within barrier and other animal housing areas to minimize the researcher's regulatory burden associated with these inspections. Many institutions incorporate the evaluations of these animal use areas into their postapproval monitoring (PAM). Many institutions require researchers to provide scientific rationale for maintaining their own, independent areas.

9.5.1 Nonsurgical Procedures

There are many nonsurgical procedures in animals in which anesthesia is used solely as a method of chemical restraint. These procedures most commonly include blood collection and imaging.

9.5.2 Surgical Procedures

- The USDA broadly classifies surgical procedures as minor or major, a classification also used by the *Guide*.

9.5.2.1 AWR

- §2.31(ix) "Activities that involve surgery include appropriate provision for pre-operative and post-operative care of the animals in accordance with established veterinary medical and nursing practices. All survival surgery will be performed using aseptic procedures, including surgical gloves, masks, sterile instruments, and aseptic techniques."

9.5.2.2 *Guide*

- Defers the definition of major versus minor procedures to that stated in the AWA but goes on to state that "Some procedures characterized as minor may induce substantial post- procedural pain or impairment."

- The IACUC plays a critical role, which begins with the protocol review. The *Guide* highlights appropriate anesthesia, analgesia, conducting surgical procedures, postsurgical care/observations.

- The *Guide* provides some direction (p. 32) regarding those researchers conducting field research where the animals are undergoing anesthesia and surgery.

Chapter 4 (Veterinary Care) and Chapter 5 (Physical Plant), among other areas of the *Guide*, provide direction regarding surgical procedures and surgical facilities.

9.5.3 Major Procedure

- AWR defines a major procedure as "any surgical intervention that penetrates and exposes a body cavity or any procedure which produces permanent impairment of physical or physiological functions."

- The *Guide* (p. 30) states, "Some procedures characterized as minor may induce substantial post-procedural pain or impairment and should similarly be scientifically justified if performed more than once in a single animal."

- Some IACUCs classify laparoscopic procedures as a minor procedure; while the procedure penetrates a body cavity, they did not consider it "exposed." Laproscopic and other "-scopic" procedures in humans are associated with less pain and infection risk.

- IACUCs should offer researchers guidance on the institutional meaning of "permanent impairment of physical or physiological functions."
- Many institutions require analgesia for minimum period following a major or minor procedure, unless scientifically justified.

9.5.4 Minor Procedure

- By default, anything that is not a major procedure is left as a minor procedure.
- Some IACUCs may be uncomfortable with the lack of regulatory guidance regarding minor procedures and, due to the scope of the individual research project taken as a whole (e.g., multiple punch biopsies), choose to label a procedure or set of procedures as a major procedure.

9.5.5 Multiple Survival Procedures

- AWR §2.31(x) states, "No animal will be used in more than one major operative procedure from which it is allowed to recover, unless (A) justified for scientific reasons by the principal investigator, in writing; (B) required as routine veterinary procedure or to protect the health or well-being of the animal as determined by the attending veterinarian; or (C) in other special circumstances as determined by the Administrator on an individual basis."
- The Guide (p. 30) states, "multiple surgical procedures on a single animal should be evaluated to determine their impact on the animal's well-being. Multiple major surgical procedures on a single animal are acceptable only if they are (1) included in and essential components of a single research project or protocol, (2) scientifically justified by the investigator, or (3) necessary for clinical reasons....Cost savings alone is not an adequate reason for performing multiple major survival surgical procedures."

9.5.6 Nonsurvival Surgical Procedures

- Rodents may undergo nonsurvival surgical procedures as part of an IACUC-approved research proposal. The key element is that the animals are appropriately anesthetized throughout the procedure and that the animals do not regain consciousness.

- Each institution should develop an appropriate policy related to the type and scope of the research conducted. Development of a policy should include discussions with occupational/ environmental health and safety professionals in addition to veterinary staff, in the event that the animals are exposed to a compound or other agent prior to the nonsurvival surgical procedure (e.g., infectious agent, bromodeoxyuridine exposure) or during the nonsurvival surgical procedure (e.g., urethane anesthesia, paraformaldehyde perfusion) or if another relevant consideration is needed.

9.6 location

- AWR: §2.31(ix) "all surgery on rodents do not require a dedicated facility, but must be performed using aseptic procedures."
- *Guide*: pp. 116–123 and other areas
  - Permits aseptic surgery in dedicated facilities or spaces.
  - States that surgical facilities are in low traffic areas that are maintained and operated in ways to ensure cleanliness and that "it is imperative that the room [used for surgery] be returned to an appropriate level of hygiene before its use [as a surgical area]"
    - A rodent surgical area can be a room or portion of a room that is easily sanitized and not used for any other purposes during the time of surgery.
    - Dedicated rodent surgical facilities are preferred, especially for immunocompromised rodents or those receiving exposed implants (e.g., headcaps).
    - PIs requesting a survival surgical area in their laboratory are expected to provide scientific justification.
    - Investigator-maintained surgical areas require periodic IACUC inspection.
    - Investigator-maintained surgical areas may expose individuals not otherwise on IACUC-approved protocols to allergens or other OHS/EHS matters. Those

exposed would include sanitation and maintenance staff, laboratory personnel, and laboratory visitors.

- Surgical areas must provide the following areas, with appropriate separation:
    - Patient preparation
    - Surgeon preparation
    - Instrument preparation
    - Surgical area
    - Patient recovery
- Particular attention must be focused on what happens to animals (specifically rats and mice) following surgical procedures. Many facilities have made a heavy investment in preventing the introduction of deleterious agents into an animal facility. Animals having surgical procedures in areas outside vivarium procedure rooms may not have the HEPA-filtered air or other amenities preventing deleterious agent introduction. Although the surgery, or other procedure, has been performed aseptically or appropriately, animals should be maintained in a "step-down" area or "return room." Special care must be taken with immuno-compromised, breeding, or immunovague animals.
- Animals maintained outside the vivarium (for post-operative evaluation and other reasons) must meet all the requirements of being housed in the dedicated vivarium, including IACUC inspection, regular cage changes, light cycles, appropriate diet, and so on, with the corresponding records. Intensive postoperative evaluation and assessment must be properly documented, especially when animals are maintained outside a dedicated vivarium.

9.7 key personnel interactions and roles

9.7.1 Institutional Animal Care and Use Committee

US statutory requirement supported by the Guide *(p. 13* Guide)

- "responsible for assessment and oversight of the institution's Program components and facilities. It should have sufficient authority and resources (e.g., staff, training, computers and related equipment) to fulfill this responsibility." (p. 14 *Guide*).

- "Recurrent or significant problems involving experimental animal health should be communicated to the IACUC, and all treatments and out- comes should be documented" (p. 114 *Guide*).

9.7.2 Composition

- AWR §2.31(b)(2): "The Committee shall be composed of a Chairman and at least two additional members of the Committee: At least one shall be a Doctor of Veterinary Medicine, with training or experience in laboratory animal science and medicine, who has direct or delegated program responsibility for activities involving animals at the research facility; At least one shall not be affiliated in any way with the facility other than as a member of the Committee, and shall not be a member of the immediate family of a person who is affiliated with the facility...such person will provide representation for general community interests in the proper care and treatment of animals...not more than three members shall be from the same administrative unit of the facility."

- *Guide*, pp. 24–25: "Committee membership includes...a Doctor of Veterinary Medicine either certified...or with training and experience in laboratory animal science and medicine...at least one practicing scientist...at least one member from a nonscientific background...at least one public member to represent general community interests in the proper care and use of animals...no more than three voting members should be associated with a single administrative unit."

9.7.3 Roles

- The IACUC must develop, with the attending veterinarian (AV), concise written guidelines for the research staff as it pertains to anesthesia and anesthetic-related procedures.
- The IACUC must encourage the research staff to work closely with the AV and other relevant parties (e.g., EHS, OHS) to develop precise research proposals that will be submitted to the IACUC.

- While the AV and veterinary staff may be "heavy lifters" for the IACUC, the veterinary staff must not drive the committee.
- The veterinary staff can play a crucial role in developing written anesthesia, analgesia, and sedation policies and guidelines facilitating research staff that are completing IACUC protocols.

9.8 compliance tools

9.8.1 Animal Care and Use Protocol

- The PI completes an animal care and use protocol (ACUP), which provides information for the IACUC to review about their research paradigm. The ACUP will vary, sometimes drastically, from institution to institution, as does the general review process and procedures. To facilitate the ACUP submission and review process, many institutions have developed an online protocol tracking system to facilitate and expedite the approval, implementation, and activation processes.
- Researchers who may be moving to a new institution should consider meeting with the IACUC staff and/or chair, in addition to the AV, early in the recruitment process to fully understand the similarities and differences between the two institutions.

9.8.1.1 Institutional guidelines or policies

Institutions develop and adopt various guidelines or policies to facilitate PIs in completing the ACUP (e.g., anesthesia, analgesia, and tranquilizer guidelines, blood collection guidelines, compound administration guidelines). These guidelines and policies do change periodically, research staff are encouraged to be aware of the various guidelines or policies generated by the institution. Likewise, *de novo* guidelines and policies may be generated by the IACUC.

9.8.1.2 Pre-committee review

- Many institutions incorporate a *pre-committee review* (PCR) before the protocol is formally submitted to the IACUC so that pertinent and timely IACUC-related issues can be addressed in addition to matters relating to animal husbandry and veterinary care. Some institutions choose to have the review completed by the AV or their veterinary designee, others by an IACUC-member scientist, and a very few have the review completed by a nonscientist/community member, IACUC-member scientist, *and* a veterinarian. The latter PCR provides a comprehensive review by the three main IACUC-mandated member groups.

- The PCR should be documented and completed in a short period (e.g., 2–3 business days). The PCR can be used as a method to identify protocols that may be eligible for "designated member review" as discussed in the AWRs, §2.31.

9.8.2 Record Keeping

Record keeping is a key scientific method component. Below are examples of the various records that may be beneficial, especially if there is an untoward outcome. Depending on the species, procedure, IACUC requirements, scientific necessity, and so on, additional details regarding the experimental paradigm may be needed. In the case of good laboratory practices, significantly more detail may be required and the resulting records will be closely scrutinized.

All scientists should be aware that the records they will be generating may be subject to the Freedom of Information Act (FOIA). Great care must be taken to generate accurate, thoughtful records whenever using animals for research, teaching, and/or testing purposes, especially records that may include audio, video, still images, or other such unique records. There are instances where these records have, through animal rights organizations, been FOIA-ed, reproduced, or otherwise made available to the general public. Institutions have obligations when providing such records. Institutions are strongly encouraged to seek legal advice *and* consult with knowledgeable individuals regarding the handling of such a request and carefully redacting appropriate details.

The various medical records below must be available to the AV and his/her representative. For many reasons, it is most beneficial for the records to be immediately available to the AV and/or the veterinary staff. If an animal is found doing poorly and the procedure, anesthesia, or postprocedure records are unavailable or there are inappropriate entries, then this may cause great regulatory concern. A responsible team approach, clearly articulated and supported by the IO and implemented by the IACUC, AV, and the research staff, will benefit all research team components.

Separate procedure, anesthesia, and postprocedure records may be maintained, or it may be possible to combine them into a single record. It is generally accepted that multiple rodent surgeries may be accomplished in a single day, and it may be feasible to combine a reasonable number of surgical procedures into a single record for rats and mice; other covered rodent species may need individual records. These matters must be clearly addressed as part of the IACUC guidelines. The institution's record-keeping templates should be reviewed periodically to facilitate investigator compliance.

Record availability is an important issue surrounding the various records that must be generated. Generally, these records should be given to the veterinary staff because the various regulatory bodies or voluntary compliance organizations approach the veterinary staff for this information. If kept by the researcher, the records may not be immediately available to regulatory officers or voluntary compliance site visitors, which may have negative consequences. The IACUC guidelines should address this matter specifically and clearly.

9.8.2.1 Procedure record

This may be used for a surgical or a nonsurgical procedure requiring anesthesia. The procedure record provides essential contact information should procedure-related complications develop.

9.8.2.2 Anesthesia record

The anesthesia record is important in case there are anesthesia-related complications as it can give important information when the complication occurred. Considering that various physiologic parameters are easily monitored in rodents, it is important that due consideration be given to monitoring these parameters. Although the equipment may be expensive, this adds additional merit to using a centralized, well-stocked, dedicated surgical facility rather than a given researcher's laboratory.

9.8.2.3 Postprocedure record

The postprocedure record provides details on anesthetic emergence and patient recovery from the procedure. It is important that careful records are taken in case there are unexplained events, which may include animal death. Careful records may identify when and how an animal's death or other untoward event occurred; therefore, the animal(s) should be monitored during recovery every 15 min, more frequently if complications have been observed on previous procedures or if a patient had any difficulties at any time throughout the procedure.

9.8.2.4 Combined anesthesia, surgery, and postprocedure record

The combined one-page record (front and back) is designed to monitor a few animals; the resulting records are placed in a binder. Generally, it is important that the records are held where the AV, PI, and veterinary staff have access for regulatory reasons. Copies can be made of the original records for research purposes. Examples of anesthesia records are in "Management of Anesthesia" (Chapter 4).

9.8.2.5 Controlled substance records

Controlled substances are commonly used to provide anesthesia and/or analgesia to rodents. Whenever controlled substances are used, they must be used according to the pertinent rules and regulation of the institution, state, and federal government. These rules and regulations require licenses, extensive record keeping, and proper storage practices. Individuals using controlled substances and the IACUCs must review their storage and documentation requirements periodically to ensure compliant practices are followed.

The following, at a minimum, may be required when using controlled substances:

- Requisitions, invoices, and packing slips
- DEA Certificate of Registration 223
- State controlled substances license/permit
- DEA Form 223: Controlled Substances Registration Certificate
- DEA Form 222: US Official Order Forms—Schedules I & II
- Inventory records
- Use and administration records
- Transfer records
- Disposal records
- DEA Form 41: Registrant Record of Controlled Substances Destroyed
- DEA Form 106: Report of Loss or Theft of Controlled Substances

9.8.2.6 Necropsy record

It is highly recommended to consult a veterinarian or a veterinary pathologist as part of the experimental design to optimize the pathology-related needs. The veterinarian or veterinary pathologist is especially important if an infectious disease process is suspected. The veterinarian or veterinary pathologist can work together with the researcher to ensure that the samples collected are appropriate for both research and diagnostic purposes. Rodent tissues undergo postmortem changes (PMC) rapidly; therefore it is best to euthanize rodents and perform a necropsy. Carcass refrigeration will slow PMC, not prevent them. Freezing the carcass is generally unacceptable because of the resulting tissue architecture destruction.

Each institution should develop a method for documenting necropsy findings, comprised of a complete clinical and experimental history. Electronic record databases facilitate providing the prosector/pathologist with the clinical and experimental background (Sharp and Reed 2001, 2002; Sharp et al. 2005). Documentation must include records of any samples (gross or histopathology) or photographs taken for further evaluation; such information will be compiled for each case and included, as needed, in a final pathology report.

9.8.3 Cage Cards

The cage card is an important tool too. Facilities may develop special treatment or other cards to (1) identify postoperative or postprocedure animals and (2) to help ensure that patient observations, analgesics, or antibiotics are appropriately and routinely administered as outlined in the IACUC-approved protocol. In the case of animals needing other special requirements (e.g., water, diet), these IACUC-approved requirements should also be indicated on the cage card.

Following some surgical/nonsurgical anesthetic procedures, a special diet/water may be part of the research paradigm. At many institutions, cage cards include the inclusive dates for the special diet/water and responsible contact details should the food/water run low/out. Research staff should understand that should food/diet run out, the default is that animals will be fed or given water accordingly. It is the research staff's responsibility to ensure that any/all contact details provided are accurate and up to date.

9.8.4 Hazard Use and Identification

When various hazards (e.g., viral vectors, toxins, radiation) are used as part of an animal research protocol involving animal anesthesia, they must be assessed and evaluated by appropriately trained professionals and managed by all individuals with animal contact. IACUC must be aware of the research-related hazards. As a part of the hazard assessment, evaluation, and management, appropriate signage and training for each hazard is required. Some institutions may find an agent summary sheet helpful in that it provides a concise document outlining a given hazard, where the animal can be housed, and the expected husbandry practices expected (e.g., personal protection equipment [PPE], disposable caging) to mitigate personnel hazard exposure. The agent summary sheet can be incorporated into the institution's disaster plan by providing a hazard summary to first responders and other emergency personnel in an appropriate location.

references

AAALAC International. [Internet]. Categories of Accreditation [Cited 31 May 2015] Available at: www.aaalac.org/accreditation/categories.cfm.

AAALAC International. 2008. The exit briefing: Its purpose, what to expect and how to respond. *AAALAC i-Brief,* fall 2008.

Anderson, L. C. 2007. Institutional and IACUC responsibilities for animal care and use education and training programs. *ILAR J* 48:90–95.

American Veterinary Medical Association. [Internet]. 2013. AVMA guidelines on euthanasia. [Cited 22 March 2015]. Available at: www.avma.org.

Code of Federal Regulations. 2008. Animal Welfare. Federal Register Title 9 Subchapter A.

Council for the International Organization of Medical Sciences and International Council for Laboratory Animal Science (CIOMS/ICLAS). 2012. International Guiding Principles for Biomedical Research Involving Animals. CIOMS/ICLAS.

Food and Drug Administration. 1978. Nonclinical Laboratory Studies. Good Laboratory Practice Regulations. Federal Register, Part 2, 59986–60025.

Foshay, W. R. and P. T. Tinkey. 2007. Evaluating the effectiveness of training strategies: Performance goals and testing. *ILAR J* 48:156–162.

Greene, M. E., M. E. Pitts, and M. L. James. 2007. Training strategies for institutional animal care and use committee (IACUC) members and the institutional official (IO). *ILAR J* 48:131–142.

IRAC [Interagency Research Animal Committee]. 1985. U.S. Government Principles for Utilization and Care of Vertebrate Animals Used in Testing, Research, and Training. Federal Register, May 20, 1985. Washington, DC: Office of Science and Technology Policy. Available at http://oacu.od.nih.gov/regs/USGovtPrncpl.htm; Accessed May 10, 2010.

NIH (National Institutes of Health). 2015. Public Health Service Policy on Humane Care and Use of Laboratory Animals. Bethesda, MD: Office of Laboratory Animal Welfare, US Department of Health and Human Services, NIH.

NRC (National Research Council). [Internet]. *Guidelines for the Care and Use of Mammals in Neuroscience and Behavioral Research.* Washington, DC: National Academies Press. Available at: https://grants.nih.gov/grants/olaw/National_Academies_Guidelines_for_Use_and_Care.pdf.

NRC. 1991. *Education and Training in the Care and Use of Laboratory Animals: A Guide for Developing Institutional Programs.* Washington, DC: National Academy Press.

NRC. 1997. *Occupational Health and Safety in the Care and Use of Research Animals.* Washington, DC: National Academies Press.

NRC. 2011. *Guide for the Care and Use of Laboratory Animals*, 8th edn. Washington, DC: National Academies Press.

Office of Laboratory Animal Welfare. [Internet]. Animal Welfare Assurance for Domestic Institutions. [Cited 31 May 2015]. Available at: http://grants.nih.gov/grants/olaw/sampledoc/assur.htm.

Rosenblatt, C. and P. Sharp. 2015. Introduction to the IACUC: Its purpose and function. In *Care and Feeding of an IACUC. The Organisation and Management of an Institutional Animal Care and Use Facility*, 2nd edn. Edited by W. K. Petrie and S. J. Wallace. Boca Raton, FL: CRC Press.

Russell, W. M. S. and Burch, R. L. 1959. *The Principles of Humane Experimental Technique.* London, UK: Methuen.

Sharp, P. E. and M. D. Reed. 2001. Use of a Palm OS® as a portable database and clinical management tool. *Contemp Top Lab Anim* 40(4):94.

Sharp, P. E. and M. D. Reed. 2002. Palm OS applications facilitating biomedical research. *Contemp Top Lab Anim* 41(4):86.

Sharp, P. E. et al. 2005. Developing and implementing a real-time web-based information portal for biomedical research facilities. *Contemp Top Lab Anim* 44(4):89.

Sharp, P. E. and J. S. Villano. 2013. *The Laboratory Rat*, 2nd edn, edited by P. E. Sharp and J. S. Villano. Boca Raton, FL: CRC Press.

Sikes, R. S., Gannon, W. L., and the Animal Care and Use Committee of the American Society of Mammalogists. [Internet] 2011. Guidelines of the American Society of Mammalogists for the use of wild mammals in research. [Cited 22 March 2015]. Available at: http://www.mammalsociety.org/uploads/Sikes%20et%20al%202011.pdf.

US Department of Agriculture (USDA). 2013. *Animal Welfare Act and Animal Welfare Regulations: Animal Care Blue Book*. Code of Federal Regulations (CFR), Title 9, Chapter 1, Subchapter A, Parts 1–4.

US Department of Agriculture (USDA). 2014. Animal Care Resources Guide Policies. [Cited 31 May 2015]. Available at: http://www.aphis.usda.gov/animal_welfare/downloads/Animal%20Care%20Policy%20Manual.pdf.

US Department of Justice, Drug Enforcement Administration, Office of Diversion Control. 2013. Controlled Substances Act and Regulations. Title 21, Chapter 13. [Cited 31 May 2015]. Available at: http://www.deadiversion.usdoj.gov/21cfr/cfr/index.html.

US Department of Justice. Freedom of Information Act. Title 5, Chapter 5, Subchapter II. [Cited 21 June 2015]. Available at: https://www.law.cornell.edu/uscode/text/5/part-I/chapter-5/subchapter-II.

index